Eco-efficient Rendering Mortars

Woodhead Publishing Series in Civil and Structural Engineering

Eco-efficient Rendering Mortars

Use of Recycled Materials

Catarina Brazão Farinha
CERIS, Civil Engineering Research and Innovation for Sustainability, Instituto Superior Técnico, University of Lisbon, Portugal

Jorge de Brito
Professor of Civil Engineering, Department of Civil Engineering, Architecture and Georesources, Instituto Superior Técnico, University of Lisbon, Portugal

Rosário Veiga
National Laboratory for Civil Engineering, Buildings Department, Buildings Coating Finishes and Thermal Insulation Unit, Portugal

ELSEVIER

WP
WOODHEAD PUBLISHING
An imprint of Elsevier
elsevier.com/books-and-journals

Woodhead Publishing is an imprint of Elsevier
The Officers' Mess Business Centre, Royston Road, Duxford, CB22 4QH, United Kingdom
50 Hampshire Street, 5th Floor, Cambridge, MA 02139, United States
The Boulevard, Langford Lane, Kidlington, OX5 1GB, United Kingdom

British Library Cataloguing-in-Publication Data
A catalogue record for this book is available from the British Library

Library of Congress Cataloging-in-Publication Data
A catalog record for this book is available from the Library of Congress

ISBN: 978-0-12-818494-3

For Information on all Woodhead publications visit our website at https://www.elsevier.com/books-and-journals

Publisher: Matthew Deans
Acquisitions Editor: Glyn Jones
Editorial Project Manager: Gabriela D. Capille
Production Project Manager: Vijayaraj Purushothaman
Cover Designer: Miles Hitchen

Typeset by Aptara, New Delhi, India

Contents

Mortars and sustainability

1

Mortars are construction materials composed of three main constituents: aggregates, binders, and water. There are other constituents that can be added to improve or modify the mortars' properties, according to their use, namely: admixtures, such as plasticizers, water repellents, air entrainers, or others; and additions, such as pozzolans, fillers, and fibers.

Bedding, render, screed, reinforcement, or grout are examples of different functions of mortars. Each mortar application has specific uses and therefore specific requirements. The proportion between the aggregates and the binder, as well as the water content, depends on the use. The type of binder is also chosen according to the application.

Bedding (illustrated in Figs. 1.1 and 1.2) and reinforcement mortars have the highest requirements in terms of being mechanically resistant. These mortars need considerably higher strength than the remaining ones. In this sense, they are usually richer in binder content. The binder depends on the applications. A proper compatibility between the binder and the wall is needed so that the elements function as a single element and withstand the stresses that they are submitted to. Grout mortars (illustrated in Fig. 1.3), on the other hand, have totally different functions. This type of mortar is used to fill empty spaces in masonry walls, conferring a higher compacity and cohesion to those elements. In this sense, the mortars should be fluid and plastic to be able to fulfill these spaces. Screeds, contrary to grouts, are mortars used as floors or to level horizontal elements. Render mortars (illustrated in Fig. 1.4) need very different types of characteristics. These mortars are the ones that need the lowest binder content. Common proportions between binder and aggregates can vary between 1:2 and 1:6, depending on the binder and the function to fulfill. The binder used also depends on the substrate. For more absorbents substrates that can absorb water by capillarity, the render mortar chosen to be applied should have specific characteristics. Namely, the mortars should be porous to promote drying. In those cases, air-lime or hydraulic lime binders must be chosen. Cement, on the other hand, is used to confer strength to the render, being able to protect the substrate from external actions, such as blows. This type of binder also presents a higher protection to water penetration as it absorbs a lower amount of water than the other types of binders. Thus, the type of binder chosen should take in account the intended functions and should be compatible with the substrate. This compatibility is essential for the mortar to accomplish their function of protecting the substrate.

Independently of the type of mortar, these construction materials use natural resources and are responsible for several environmental impacts.

The extraction of natural aggregates from rivers or sandbanks or the exploration of stone aggregates (such as limestone or others) using explosives are responsible for some environmental impacts. The extraction of these aggregates has consequences on

Eco-efficient Rendering Mortars. DOI: 10.1016/C2018-0-04147-X

Fig. 1.1 Illustration of an air lime–based bedding mortar, in a stone masonry wall.

the flora and fauna of the surrounding environment as well as on polluting the soil and air. Furthermore, these resources are limited and take thousands of years to be replaced again in nature. Cutting down on this extraction is an important issue that should be taken into account in the development of new products.

Binders are highly polluting materials. Usually, cement is the most used binder nowadays, due to its strength and price. Cement is in fact the most used material worldwide, after water [1] [2]. This construction material consumes a great amount

Fig. 1.2 Illustration of a cement-based bedding mortar (in a ceramic brick wall).

Fig. 1.3 Grout mortar injection illustration.

Fig. 1.4 Render mortar illustration.

of raw materials, such as clay and limestone, as well as energy, while also releasing high quantities of carbon dioxide [3] and other gases into the atmosphere, during calcination. According to the European Cement Association (CEMBUREAU), a ton of cement produced releases 738 kg of carbon dioxide into the atmosphere [4]. It is necessary to improve the cement production to avoid the emissions of gases to atmosphere and to reduce other types of environmental impacts.

In addition to the environmental impacts produced by the binder's industry, by the natural resources' extraction industry, and by the chemical industry of admixtures, there are enormous amounts of wastes daily produced by many industries and by the society, contributing also to other environmental impacts. The high daily production of wastes is no longer compatible with the present world, and waste management is a current issue in Society. Without viable solutions of reutilization or recycling, wastes are continuously sent to landfill and this procedure has several associated issues. A great part of these wastes takes thousands of years to degrade. Thus, with the increase of waste production, landfills are starting to be overloaded [5]. Adding to space scarcity, waste disposal in landfill causes environmental impacts such as soil, air, and water pollution [5] [6]. To avoid landfill deposition, it is necessary to find viable solutions for the wastes' reutilization, recycling, or another valorization. These solutions will give rise to new uses, extending the life cycle of the wastes and delaying their deposition.

The construction sector is responsible for a great amount of wastes produced in the European Union and construction and demolition waste achieved almost a third of the total wastes generated in these countries. This waste includes concrete, ceramics, mortars, stone, gypsum, wood, glass, metals, and others. In Fig. 1.5, an example of construction and demolition waste is illustrated. Simultaneously with a great production of wastes, the construction industry continues to use virgin resources to produce new materials.

Fig. 1.5 Construction and demolition waste's illustration.

The incorporation of wastes in construction materials replacing sand or cement contributes to a dual environmental benefit: it reduces the amount of wastes that are sent to landfill and reduces the amount of natural resources extracted. The encapsulation of these wastes in hydraulic mortars or concrete can be a solution for these wastes' management. The replacement of natural aggregates or cement with wastes reduces both the environmental impacts produced by the raw materials' extraction, such as emissions of CO_2 into the atmosphere, and the amount of wastes sent to landfill, reducing some environmental problems such as respiratory human health problems or pollution of the ground water.

Over the years, some studies about the incorporation of wastes in hydraulic materials can be found [1], [7-51]. Concrete is maybe the most researched construction material over the last years. The incorporation of wastes in this material has been deeply investigated. There are also some studies about the incorporation of wastes in mortars, namely masonry or reinforcement mortars [9-11], [28], [29], [40], [47], [52-67]. The incorporation of wastes in render mortars is still seldom investigated. There are some studies that can be found, namely incorporating fiber wastes, mostly natural ones [68-74]. However, the universe of the wastes is diverse and the potential of these wastes in render mortars is far from being thoroughly investigated.

This book is dedicated to render mortars and to wastes incorporated in them. In this document, the main functional requirements of this type of mortars are described and some examples of laboratory tests that can be performed to evaluate their performance are identified. Wastes are another theme that is analyzed in this book. The treatments that wastes can go through before their incorporation in render mortars and how these treatments may influence the wastes' particles characteristics are also analyzed in this book.

Afterwards, the incorporation of wastes in render mortars and how they may affect their properties is also presented. The influence of these mortars in the fresh state, mechanical behavior, and water behavior is discussed. For this analysis, examples of wastes already tested by several authors were added. The influence of wastes in workability, bulk density, and water retention, are some of the properties analyzed in the fresh state. Concerning the mechanical behavior, compressive, flexural, adherence, and impact are examples of properties analyzed in this book. Deformability and microcracking pattern are issues also included in mechanical behavior. Cracking is one of the most frequent anomalies that can occur in render mortars and is one of the main reasons for their degradation. Concerning water behavior, the influence that the wastes have on the mortars' water absorption is also analyzed, by capillarity or under pressure, and their ability to release the amount of water contained in their pores after this absorption. The influence of wastes on these characteristics is essential to analyze their viability in terms of technical behavior. To evaluate the global performance of render mortars with wastes' addition, it is also necessary to evaluate their influence on mortars' durability. In this sense, a chapter of the book is dedicated to this subject.

Finally and after a careful analysis of the mortars' technical performance, an analysis of their environmental performance is also presented. Examples of studies about the life cycle assessment of mortars with wastes incorporation, in substitution of other constituents as a natural aggregate or binder, are presented. As described in the dedicated chapter, the wastes provide great improvements in terms of environmental performance, reducing the mortars' impacts in several environmental categories.

In this book, some examples of eco-efficient mortars are presented. These mortars are sustainable materials that incorporate wastes presently with no reuse or recycling solutions, and that would be sent to landfill and would take thousands of years to be degraded. These wastes, replacing the natural aggregate or the binder of the mortars, are also contributing for a reduction of several environmental impacts. Simultaneously, the incorporation of wastes can be responsible for improving the mortars' technical performance.

Rendering mortars

2

2.1 Introduction

The skin of external walls can be made of different materials/systems, such as rendering mortars, ceramic surfaces, ventilated façades, and external thermal insulation composite systems. The choice of the envelope solution for each construction is mainly based on historical practice and varies from country to country. These solutions can have specific complementary functions such as thermal insulation. However, all have a primary similar function that is the protection the substrate. They are in fact often sacrificial layers that intend to protect the substrate from external actions, such as wind, rain, or impacts, preventing its degradation.

2.2 Technology

Rendering mortars are composed of a homogeneous mixture of one or more binders, aggregates, and water. The mortars can also contain admixtures and additions. Aggregates can be likened to the skeleton of the mortars, the binder is the constituent that agglomerates the solid particles and water is the element that confers workability and, in the case of the hydraulic binders, is responsible for the hydraulic chemical reactions.

Nowadays, aggregates can come from different origins: natural sources, artificial production, or wastes, that are rejects from the population or from other industries. Natural aggregates usually come from rocks or sediments. In locals near coastlines or rivers, natural aggregates usually come from riverbeds or sandbars, which sometimes is not possible or economically viable. In those cases, aggregates can be obtained by crushing rock blocks. The local of extraction defines the nature of the aggregates, for example they can be siliceous, calcareous, argillaceous, granitic, basaltic, or others [75].

Artificial aggregates are aggregates produced purposely with that aim. There are several light aggregates artificially produced and incorporated in mortars or concrete, such as aggregates of expanded clay, expanded slate, slag, or others [76]. The use of wastes as aggregates has been investigated over the years. Some wastes can be reused or recycled and incorporated in mortars as aggregates, as will be presented next. Wastes can be organic or inorganic. The origin of the aggregate influences their chemical and physical characteristics such as the shape and roughness. Independent of the origin, the grain size of aggregates varies between tens of micrometers and 5 mm [77]. The grain distribution is given by different ranges of grain sizes. A proper grain distribution gives origin to a more compact mix.

Eco-efficient Rendering Mortars. DOI: 10.1016/C2018-0-04147-X

In rendering mortars, the most usual binders are cement, air lime, and hydraulic lime. A single one or a combination of two can be used. Cement should only be used in new construction, while lime and hydraulic lime are more often used in rehabilitation of old buildings [78]. The different uses are related with the compatibility with the substrates [78].

In rendering mortars, the ratio between binder and aggregates is usually considered in volume and varies according to the binder used. In air lime mortars the ratio used in render varies between 1:2 and 1:3 [78, 79]. In cement-based mortars the ratio varies between 1:3 and 1:6 [80, 81]. The water to binder (w/b) ratio is near one in cement mortars and increases in the case of aerial lime mortars.

Admixtures are chemical products used in small contents in renders to modify some characteristics of renders. Additions are products, usually mineral, added to rendering mortars to improve some properties of renders [82]. Plasticizers, air entrainers, and waterproofing products are examples of admixtures used in renders [77]. Fibers or nets are examples of additions used in renders intending to improve the flexural strength and susceptibility to cracking of mortars.

The mixing of the different constituents that compose rendering mortars can be performed *in-situ* or in a controlled environment, as in a manufacturer's premises. Nowadays, industrial mortars in paste or in powder are often used. These mortars are prepared in a controlled environment where the quality of the products and of the processes is controlled, guaranteeing a higher level of trust in the quality of the mortars.

Renders can be composed by a single layer or by several layers. When they are composed of several layers, the first one, which is in contact with the substrate, is the spatterdash coat. This layer is usually richer in cement than the remaining ones. The function of the spatterdash coat is to homogenize the absorption of the substrate, to provide a good adhesion to the substrate, and to create an irregularity in the substrate to improve the adherence of the following layer. The base layer has a thickness of about 2 cm, variable according to the regularity of the substrate. This layer can be applied in a single application or in thin consecutive layers. This last technique can improve the cracking susceptibility and the durability of the rendering. After the base layer, the finishing layer is applied. In this layer, the aggregate is finer to create a smoother surface. After the finishing layer, a painting can be applied for decorative and complementary protection purposes.

2.3 Functional requirements

2.3.1 Workability

Workability can be considered the most important property of a mortar in the fresh state. This is the characteristic of the mortar which allows a render to be properly applied on the substrate. A mortar with proper workability can be transported without segregation, can be easily applied on the substrate, is able to even the irregularities of the substrate, remains in a plastic phase during the rendering application process, and is able to adhere to the substrate [83].

Workability is a complex concept as it is a combination of several properties, including plasticity, consistency, cohesion, and adhesion. A render should be easily applied, with a good adherence to the substrate, a proper compactness, and a good yield.

This property of the mortars depends on the mortar's constituents' physical characteristics. The shape, surface area, roughness, and water absorption are some of the characteristics of the constituents that influence the workability of mortars [84]. The water needed in the mix must be able to lubricate the surface of all the constituents. This lubrication will be responsible for a homogeneous and workable paste. The shape, superficial surface area, and roughness of the constituents will influence this amount of water [82]. More angular and rougher particles will require more water to lubricate the surface of the particles, in comparison to smooth and spherical particles. There are other properties of the constituents that influence the water needs, such as their water absorption. An absorbent constituent will absorb the mixing water, and the free water used to make a plastic paste is reduced. This reduction of the mixing water will decrease the workability of the paste.

The constituents, as a whole, and their grain size distribution in particular, also influence the workability of the paste. The aggregates are, in general, the particles with larger diameter, and their grain size distribution varies between tens of micrometers until 5 mm [77]. The proportions of grains with each diameter range are variable and are responsible for the quality of the distribution. The particles rearrangement is responsible for the compactness of the mortar. The larger particles will create voids among themselves, and the smaller particles will fill those voids decreasing the volume of the air voids of the mix. A well-distributed aggregate will positively influence the workability of the mortar.

Workability is difficult to be evaluated as it is strongly related to the workmanship and subjectivity of the craftsman [83]. However, there are some laboratory tests that can be performed.

The consistency of the mortars can be evaluated by the flow table test or by penetration tests [83]. In the flow table test, the mortar, in paste, is used to fill a truncated cone mold with a given compaction. Next, the fresh mortar is subjected to a given vibration and it flows over the table. The diameter of the mortar flow is measured in more than two distinct alignments. The flow table result gives an indication of the workability of mortars. In Fig. 2.1, the consistency by flow table test is represented.

Several penetration tests are also used to evaluate the consistency of the mortars in the fresh state. They consist in dropping an object with a given mass and shape and measure the depth of penetration of the mass into the mortar. The object can be a ball, a cone, or a cylindrical plunger with hemispherical top.

2.3.2 Water penetration

Most of the materials in a construction can be subjected to water absorption, but external walls are more exposed to water than other elements. These walls can absorb water from different sources and soil and rain are common sources of water. There are some construction anomalies that increase the exposure of walls to water availability, such as deficient drainage nets or water pipe breaking.

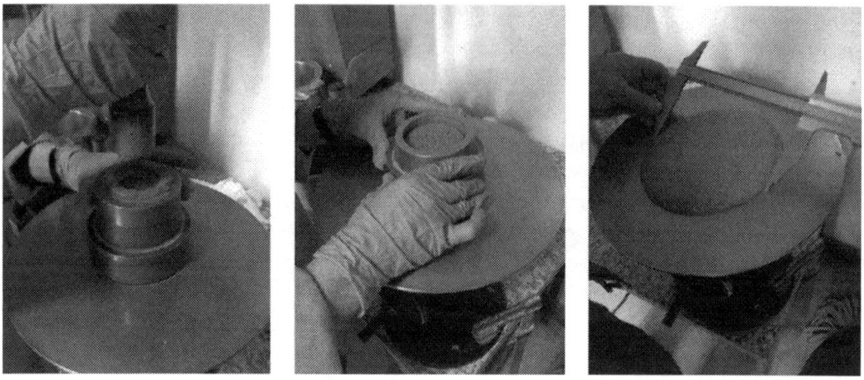

Fig. 2.1 Consistency by flow table.

Concerning the water absorption from the soil, when the wall is in contact with it, and there is no barrier to water absorption, the porous materials in contact with the soil will absorb water by capillarity. This liquid water will migrate into the interior of the wall and will evaporate in the drying zone.

The walls can also absorb water from the rain, which can occur by capillarity or under pressure, depending on the wind that blows on the surface of the external walls. Usually the finishing layer prevents the absorption of the mortars. The materials used in these layers show low permeability to liquid water, preventing the water migration into deeper layers. However, due to anomalies from these finishing layers, the water can migrate into the mortars and then limited permeability to water of the mortars is essential to prevent the migration of the water to the interior of the wall.

The water absorption of the porous materials, where mortars are included, is influenced by their internal porosity. The materials contain several pores' ranges, which can be divided into gel pores, capillary pores, and macropores. In water absorption by capillarity only the capillary pores contribute to water absorption, and the rate is influenced by the capillaries size. The larger the capillaries are the faster the water absorption is. Water absorption by capillarity is defined by the ability of a mortar to saturate due to suction forces [79]. When the water is absorbed under pressure (e.g., due to wind pressure), the macropores in addition to the capillary pores will contribute to water absorption.

The prolonged presence of water in the interior of a render can result in a decrease of mechanical strength and thermal insulation, in the transportation of salts (dissolved by liquid water) to the drying zone, in modification of the material size (shrinkage or dilatation), and can favor microcolonization [79].

The water permeability behavior of the mortars can be analyzed by the water absorption by capillarity test, the water immersion test, and the water under pressure test.

The water absorption by capillarity test consists in quantifying the mass of water absorbed when a specimen is in contact with this liquid. The mass of water absorbed, over time, per unit area of contact leads to a capillary curve, as illustrated in Fig. 2.3. In this test, the water absorption is unidirectional. For that purpose, the specimens should

Fig. 2.2 Water absorption by capillarity test.

be watertight in their lateral faces and only the base of the specimens should be in contact with water (Fig. 2.2). The specimens are weighted until they achieve a constant mass. The test should be performed in controlled conditions of temperature and humidity. With the data obtained, a curve of water absorption (kg/m^2) per time (h$^{1/2}$) can be drawn. The curve is characterized by two segments (Fig. 2.3). The first corresponds to the quick water absorption that can vary between 90 min and 24 h, depending on the mortars. The slope of this first segment is entitled capillary coefficient and is influenced by the thinner pores, within the capillary pores range. The thinner the pores are the lower the slope of the first segment of the curve is, meaning that the initial absorption is slower. The larger capillary pores are the firsts to be filled. After that, the thinner ones

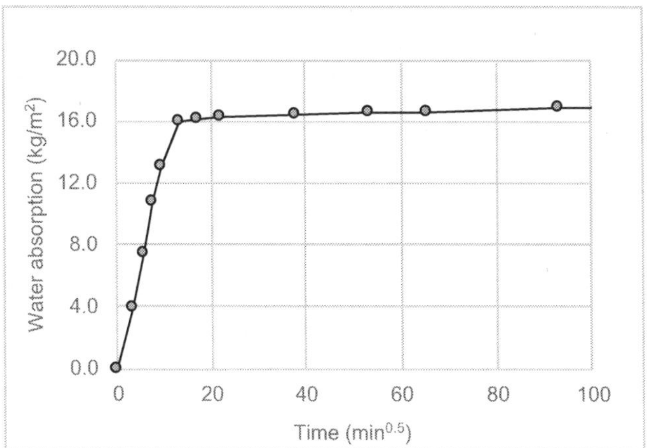

Fig. 2.3 Water absorption by capillarity typical curve.

Fig. 2.4 Example of permeability to water under pressure test (vertical substrate).

start to fill and so on until all the capillary pores are filled. At the end of the test, when the specimens achieve saturation, it is possible to quantify the total water absorption. The volume of water absorbed gives an indication of the total volume of capillary pores of the mortar.

The water permeability under pressure test consists in quantifying the water absorption of a mortar, per unit of contact area, when some pressure is applied. The pressure is achieved by a column of water. For a short period of time the water absorbed is quantified. The absorption process, depending on the tests, can last a few hours or a few days. The test should be performed under controlled conditions of temperature and humidity. In addition to the capillary pores, the macropores and cracks also contribute to this water penetration. Therefore, the combination of this test with the water absorption by capillarity test allows achieving a global idea about the water behavior of the mortars. In Figs. 2.4 and 2.5, two different tests of water absorption under pressure are represented, the first applied to a vertical substrate and the second to a horizontal substrate application.

2.3.3 Water release

Ideally, renders should have low permeability to liquid water and high permeability to vapor. The low permeability to liquid water reduces the amount of water absorbed by the mortar. The permeability to vapor allows the release of that amount of absorbed water, reducing the time that the water remains inside the render. Therefore, the water infiltrated in the walls through cracks or singularities and absorbed by the mortars should be evaporated as soon as the atmospheric conditions allow it [77]. Prolonging

Fig. 2.5 Example of permeability to water under pressure test (horizontal substrate).

the presence of water in the interior of the render gives rise to several anomalies. Biocolonization is one of the possible anomalies that can occur. In the limit, saturation of the mortar can lead to the degradation of the substrate. In this case, the render fails its main goal that is in fact the protection of the substrate.

The water vapor permeability of mortars is directly related to their internal porosity. More porous mortars release water more easily than more compact ones. All the ranges of voids contribute to water drying. In an initial phase, liquid water percolates, from the larger pores or internal microcracks, until the drying surface and then evaporation follows. At some point, the transportation of water inside the mortar until the surface starts to be in vapor, starting a mix transportation of water. Finally, transportation will be only in vapor until the mortar reaches full drying. The permeability to water vapor generally increases over time. Due to the matrix shrinkage, a degradation of the internal transition zone between the aggregates and the matrix is noticed, creating air voids between the two elements, with ageing [75]. These voids contribute positively to water drying.

To evaluate the mortars' drying behavior, some laboratory tests can be performed. Next, two tests that can evaluate this property of mortars are presented.

The water vapor permeability test consists in placing a mortar between two different humidity environments. A significant variation of humidity from the exterior and the interior of the specimen is necessary to perform the test. Then the specimen is sealed, allowing water evaporation only through the mortar. The samples are weighted in consecutive days until the mass of evaporated water is constant. Fig. 2.6 represents the water vapor permeability test. With the rate of water vapor mass released and area and width of the specimen, it is possible to measure the water vapor permeability of the mortars and determine the equivalent thickness of air.

In water vapor permeability test, the specimens are not in contact with liquid water, rather with water in a vapor state. In a real mortar's application, the specimens are

Fig. 2.6 Water vapor permeability test.

in contact with water in both the liquid and vapor states. Therefore, the water vapor permeability test has some limitations.

The drying test also evaluates the drying capability of mortars. It may be performed immediately after the water absorption by capillarity test. This is a more complete test than the water vapor permeability's as the mortars are in contact with liquid and vapor water, representing better the real application of mortars. This test consists in daily weighting the specimens that were saturated until they achieve a constant mass. The test should be performed under controlled conditions of temperature and humidity. With the data obtained in the test, the curves of evaporated water mass (per area of drying) per time or per square root of time can be drawn (Fig. 2.7).

By analyzing the curves, the different phases of water transportation can be identified [85]. The first one corresponds to liquid transportation of water from the interior pores, followed by evaporation. The second one is a mixed phase where the water is transported in the liquid and vapor forms. The last phase of transportation of water is in vapor. The slopes of the linear regression lines of each phase indicate the ability of drying. The higher the slope value for each phase, the higher the mortar drying ability. The first drying phase is quantified by the curve of evaporated water mass per time. The remaining phases are quantified by the curve of mass evaporated per square root of time.

From the curve of mass evaporated per time, another coefficient can be quantified that gives a global idea about the drying ability of the mortars: the drying index. This coefficient is represented by the area under the curve. The lower the drying index, the higher the ability of the mortars to dry [86].

2.3.4 Adhesion to the substrate

A proper adhesion, between the render and the substrate, is essential for the performance of water protection functions and is a conditioning factor of the durability of the render and the substrate [77, 79].

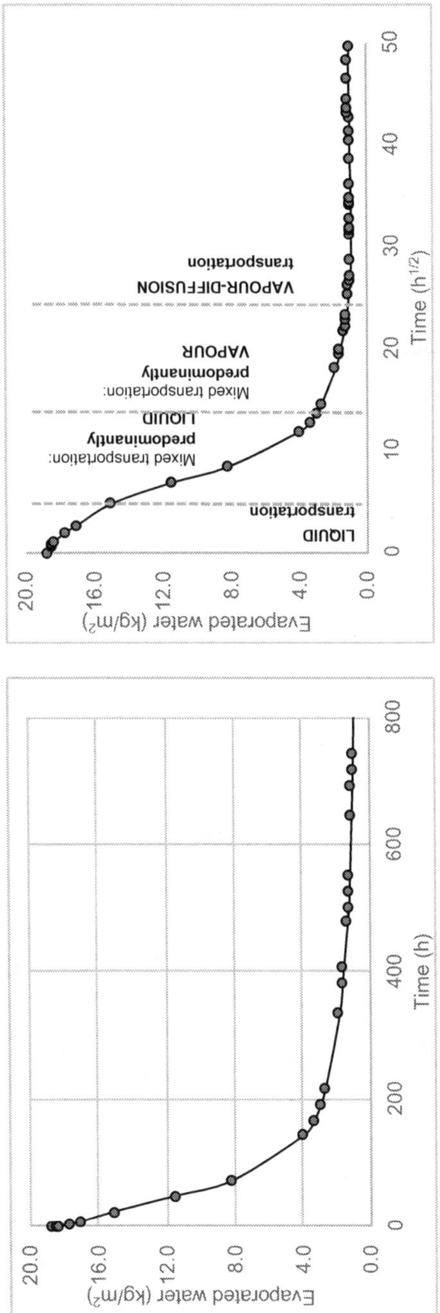

Fig. 2.7 Drying typical curves (per hour and per hour$^{1/2}$).

Adhesion to the substrate is achieved through the penetration of a mix composed of mixing water, binder, and the thinnest particles of the aggregate, present in the fresh paste of the mortar, into the pores of the substrate. This water moves thin particles that will hydrate in the interior of the substrate pores, creating a bridge between the mortar and the substrate. Adhesion depends, mainly, on the characteristics of the substrate and the mortar. The porosity and roughness of the substrate influence the adhesion. A porous and rough substrate benefits the adhesion, as it allows the fresh mortar paste to fill the rugosity and to penetrate the substrate's superficial pores. However, an overly porous substrate can absorb a great amount of mixing water and hinder the hydration reactions of the binder, creating a powdery layer that jeopardizes adhesion.

The lack of adherence can lead to the detachment of the mortar, which may be responsible for several anomalies such as cracking, increment of water absorption, or differential temperature differences. Finally, the applied mortars can be detached and exposed the substrate.

Adhesion to the substrate can be analyzed by the adherence strength, which is usually measured as the maximum tensile stress due to a direct load applied perpendicularly to the surface of the rendering mortar. From the surface inwards of the render, a core is made. This core should cut all the layers and include the interface render/substrate. Over the core a stud is glued. The tensile load is applied by specific equipment that pulls the stud from the mortar surface. The maximum tensile load, per area of contact, determines the adherence strength of the mortar. Three different fracture patterns can be obtained in the adherence strength test: adhesion fracture (fracture pattern A), cohesive fracture in the mortar (fracture pattern B), and cohesive fracture in the substrate (fracture pattern C). The fracture patterns identify the element where the fracture occurs, which can be in the interface, in the mortar, or in the substrate. The fracture occurs in the most susceptible element that composes the render. In Fig. 2.8, the three different fracture patterns that can occur in the test are represented. Only an adhesive fracture allows quantifying exactly the adherence strength between the mortar and the substrate. In the other fracture patterns, the value obtained is lower than the interface strength.

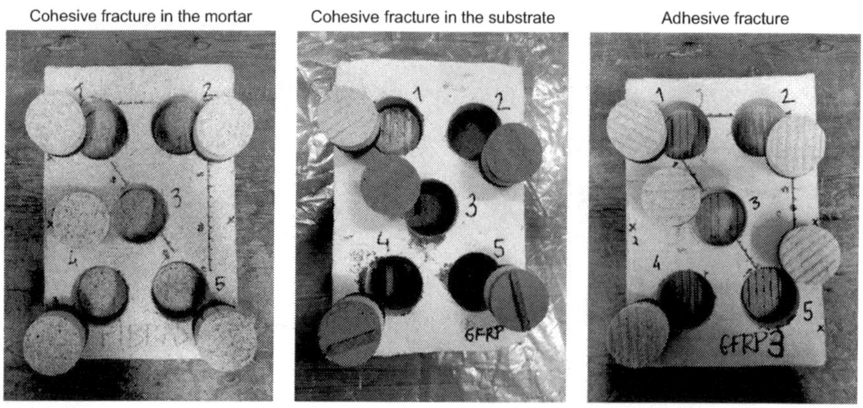

Fig. 2.8 Adherence fracture patterns.

2.3.5 Cracking

Cracking is one of the most frequent anomalies of renders. This anomaly affects the aesthetic appearance of external walls and consequently the visual appearance of buildings. If the render is not repaired, cracking can also affect the renders' durability [87]. Cracking affects the watertightness of the renders and its protection ability [77]. This anomaly creates an open path to liquid water and can also be responsible for other associated anomalies such as dissolution of salts, susceptibility to the growth of plants, or biocolonization.

Adequate design, application, and maintenance of the renders can prevent mortars' cracking.

The functional requirements of mortars, their characteristics, the characteristics of the substrate, the location, and compatibility between the different materials are essential points that have to be evaluated in mortars' design.

The location, orientation, and exposure are geographic characteristics that should be taken into account in mortar's design. In locations exposed to great variations of temperature, the choice of an adequate mortar is even more essential. Rain followed by negative temperatures is responsible for the formation of ice in the pores of mortars. This phenomenon can be responsible for the mortars' cracking. Liquid water fills these pores and, when the temperature goes down, until negative values, it is transformed into ice, occupying more space in the interior of the pores and creating stresses there. If the mortar is not adequate, it can crack during these ice-thaw cycles. In these cases, the mortars should be designed with an air entrainer admixture to increase the volume and size of the voids to reduce the stresses generated during the ice-thaw cycles. A proper flexural strength should also be taken into account to support the stresses introduced by this environmental problem. High temperatures can also be responsible for mortars' cracking. In this type of environment, the cause of cracking is the movements of the materials that compose the external wall. Different variations between the several elements lead to cracking of the surface. A similar coefficient of thermal expansion between all the elements, a lower modulus of elasticity than the substrate, and proper flexural strength are the essential characteristics that will prevent mortars' cracking.

In addition to the local of application, the properties of the mortars and substrate, as well as their compatibility, are, as mentioned, essential to prevent cracking. A chemical, physical, and mechanical compatibility between the render and the substrate is required.

A common anomaly in renders is cracking due to mortars' shrinkage. This cracking is characterized by a specific mapped pattern, which is illustrated in Fig. 2.9. The mortars, when applied on a substrate, cannot freely shrink, being restrained by the substrate. Without an adequate modulus of elasticity, the mortars can crack.

The adherence between the render mortar and the substrate is also essential to prevent renders' cracking. This bond allows both elements to function together and the tensile stresses produced to be well distributed in the mortar, thus preventing their cracking.

As in all construction materials, an inadequate application affects renders' performance. The humidity of the substrate before the mortar's application, a correct application, and the geometric dimensions designed are some of the application factors.

Fig. 2.9 Cracking pattern due to shrinkage.

As has been described there are several factors that influence render's microcracking. In this sense, it is difficult to measure these microcrackings [79].

The deformability of the mortars can be evaluated by their modulus of elasticity. This modulus can be quantified by static or dynamic methods. The static method measures the slope of the load/displacement curve in the elastic phase. However, there are other more expedite methods to evaluate the modulus of elasticity, namely the dynamic method. The dynamic modulus of elasticity can be determined by the frequency of resonance or by ultrasonic waves. The frequency of resonance allows the determination of the natural vibration frequency of each material, which gives an indication about its modulus of elasticity. To achieve the frequency of resonance, a vibration (continuous or instantaneous) is induced in a specimen. The frequency obtained in the test is used to calculate the modulus of elasticity. Ultrasound pulse velocity test consists in generating an ultrasound wave and measuring the time that it takes to go through the specimen. The velocity of the wave indicates the mortar's compacity and can be used to calculate the modulus of elasticity. The time that the wave takes from the sender's probe to the receiver's is recorded. From the data, it is possible to calculate the wave velocity. From the velocity of the wave and the characteristics of the material (Poisson coefficient and geometric dimensions of the specimen), it is possible to quantify the modulus of elasticity of the mortar.

In addition to the modulus of elasticity, shrinkage is another property that can affect mortars' microcracking [88]. There are several shrinkage tests that can be performed to evaluate the susceptibility to cracking. Plastic shrinkage can be

Fig. 2.10 Free and restrained shrinkage tests.

analyzed in laboratory. For that purpose, a mortar is applied on a substrate and drying is artificially forced. The surface of the specimen is observed, and the microcracking appearance is evaluated. The drying source can be the artificial reproduction of wind, by a wind tunnel, or fans. Free shrinkage is another test that evaluates shrinkage. In this test, the displacement is free. This test, individually, is not indicative of susceptibility to cracking. A porous mortar probably presents higher free shrinkage but it cannot microcrack. Restrained shrinkage and ring tests are other tests to evaluate microcracking with the advantage of giving quantitative results [87]. In a real application, the mortars' shrinkage is restrained by the substrate thus these tests represent better the real application conditions of the mortars to be compared with the free shrinkage test's results. In Fig. 2.10, the free and restrained shrinkage tests are presented.

2.3.6 Impact resistance

An exterior coating should be resistant to impact. This impact includes small impacts from people actions as well as small strokes or balls impacts. A render should absorb the energy produced in the impact without cracking and maintain its characteristics after small impacts.

Usually there are two methods to evaluate the impact resistance of a mortar. When a render is applied on a vertical substrate, a pendulum is used to determine the impact resistance (Fig. 2.11). In laboratory, the mortars can be applied to a substrate and falling of a mass from different heights can be indicative of their impact resistance.

The first test consists in positioning the pendulum perpendicular to the render and let a mass with a specific weight fall on the render. The diameter, depth, and cracking pattern surrounding the dent created by the impact are analyzed. These characteristics give an indication about the impact resistance of the render. The cracking pattern also allows understanding the deformability of the render.

Fig. 2.11 Impact resistance test (vertical substrate).

The impact resistance can be measured through a different approach, by the fall of a mass from different heights (Fig. 2.12). For each height, impact energy is produced. In the test, the energy levels necessary for the first crack and for failure to occur are evaluated. At both stages, the diameter and depth of the dent are evaluated. The

Fig. 2.12 Impact resistance test (horizontal substrate).

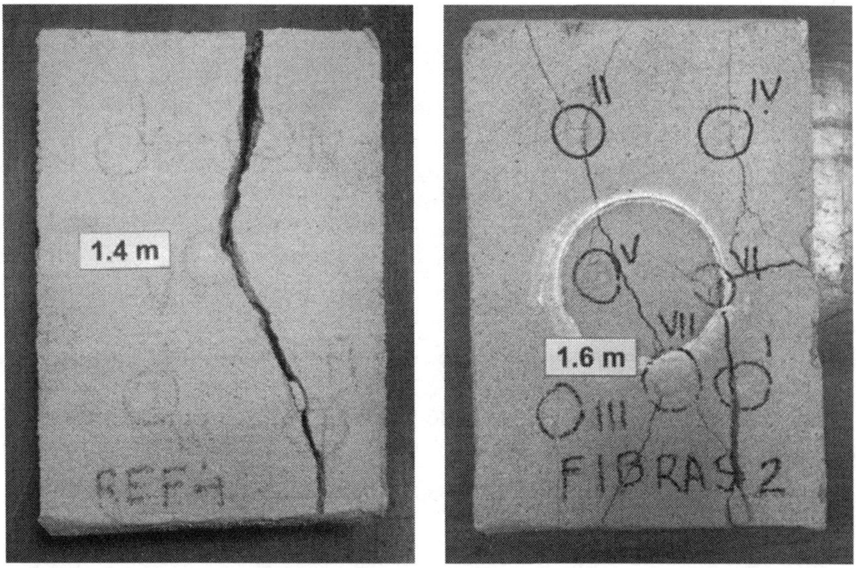

Fig. 2.13 Different failure patterns obtained in impact resistance test.

higher the height of the first crack and failure is, the higher the impact resistance of the mortar is.

At failure, the pattern of cracking is also analyzed. Brittle mortars present, in general, a single crack with large width, while more ductile mortars present thinner distributed cracks (Fig. 2.13).

2.3.7 Compatibility with the substrate

Compatibility is essential to the preservation of the substrate and the good performance of the rendering. Renders are sacrificial materials that intend to protect the substrate to extend its lifetime. Rendering solutions cannot be responsible for the degradation of the substrate; hence it is necessary to design mortars compatible with the substrate. A render mortar should be geometrically, physically, mechanically, and chemically compatible with the substrate [75].

In terms of geometry, at first, a render mortar must be able to even the surface of the substrate by filling its irregularities. The regularization of the substrate allows the application of the finishing layer that confers the render its final appearance. Also concerning geometry, a render should be compatible with the substrate in what concerns the deformability, both thermally and hygrically. The render mortar should present more deformability than the substrate, thus allowing the mortar to follow the movements of the substrate without cracking. A lower deformability of the render would generate stresses in the substrate, which could lead to the substrate's degradation.

A rendering must have low permeability to liquid water to absorb less water. However, a render mortar must be permeable to water vapor. This permeability will allow vapor exchanges between the substrate and the external environment.

These exchanges allow the substrate to breath. With a proper vapor permeability, the substrate is able to release some water that was absorbed (due to several possible sources) to the exterior, without compromising or altering their technical and chemical performance. This permeability is related to the physical compatibility between the substrate and the render.

A render has mechanical compatibility with the substrate if the mechanical strength is well adapted to the substrate. In particular, when a substrate presents low mechanical strength, the render mortar should not present higher mechanical strength to not transmit high stresses that could degrade the substrate.

A chemical compatibility is also required between the render mortar and the substrate. On the other side, the render should not contain high amounts of soluble salts in its composition. In the presence of water, salts are dissolved and can be transported to the interior of the render or even of the substrate. When the water dries, crystallization of the salts occurs, which generates stresses in the pores, leading to the loss of cohesion of the material.

2.3.8 Esthetical appearance

The esthetical appearance of a building is essential for the satisfaction of the inhabitants; therefore it cannot be ignored.

The finishing layer is mainly responsible for the esthetical appearance of the external coating of a building. However, some anomalies can occur in the render mortar that can be responsible for the degradation of the esthetical appearance of the coating. Efflorescence, crypto-florescence, cracking, biocolonization, or thermophoresis are some of these anomalies.

2.3.9 Durability

Durability is the ability of the materials to maintain the initial characteristics and performance over a given period of time. The durability of the mortars is essential for them to continue to protect the substrate.

The durability of the mortars is usually evaluated by the difference between performance before and after ageing. Ageing can be natural or artificial. Artificial ageing has some limitations, as it is not able to reproduce exactly the environmental exposure, and the period of ageing that it represents in reality is always unknown. Nevertheless, this approach is characterized by a reduced testing time. Independently of the ageing process used, the performance of the render mortars before and after ageing must be evaluated. Comparing these performances gives indication about the durability behavior of mortars.

In natural ageing, the walls should be oriented to the most unfavorable orientation. Each orientation is characterized by different variations of temperatures, different degrees of solar radiation exposure, and consequently different variations of humidity. These differences will give origin to distinct anomalies. For example, cracking is more frequent in the orientation with more incidence of solar radiation, while biocolonization is more susceptible in the orientation with less solar radiation.

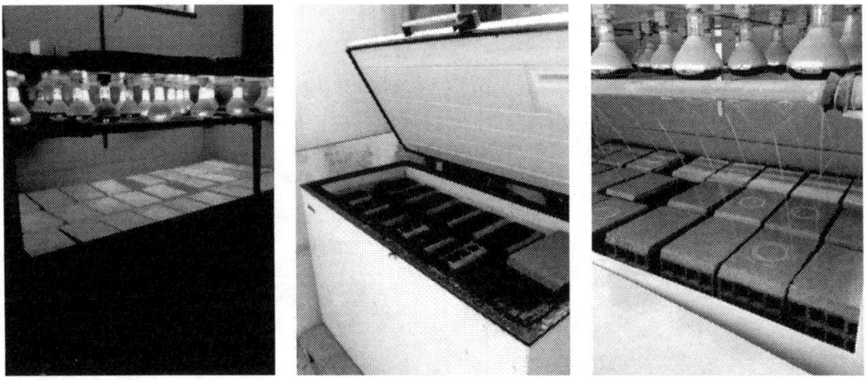

Fig. 2.14 Artificial ageing setups.

There are several methods to apply artificial ageing conditions to materials [89]. In Fig. 2.14, a possible artificial ageing process for renders is presented. In this method, the render mortars are applied to a substrate and are submitted to extreme thermal-humidity cycles [90].

Deformability, water absorption, and adherence strengths are some of the tests that are usually evaluated before and after natural and artificial ageing. The appearance of cracking after ageing is also analyzed.

The presence of biocolonization, namely molds, in the case of natural ageing can also be evaluated. The appearance of molds in coatings is a common phenomenon that affects the general appearance of buildings. Biocolonization by molds is another degradation mechanism that can occur in mortars when there is organic matter content, water availability, and a relatively high temperature. In the zones of molds development, the presence of water accumulated in these singularities is common. The prolonged exposure to water can be responsible for other anomalies, such as infiltrations or superficial stains. These molds are filamentous fungi that develop generally in moist environments rich in organic matter. In this sense, for mortars with organic matter content, the susceptibility to molds' development should be analyzed. The susceptibility to molds' colonization can be evaluated in laboratory. The test consists in placing the mortar specimens, previously sterilized, in a propitious environment for mold development. Over the specimens the inoculum of more than one mold is suspended. The degree of development of molds on the surface of the specimens is quantified weekly for a limited time, for example during the 4 weeks of exposure [91].

The organic matter affects the molds' colonization but it can also affect the fire reaction of renders. Renders and plasters are considered materials with no contribution to fire unless they present a significant amount of organic matter. If the amount of organic matter is significant, it is necessary to evaluate the fire reaction of the renders and analyze whether there is any restriction to their application in terms of height or location.

2.4 Mortars tests standardization

There are several standards to evaluate the characteristics of render mortars, enunciated in Section 2.3. In Table 2.1, the main standards are summarized, namely the European, Deutsch, English, and American Standards.

Table 2.1 Test and methods to evaluate the functional requirements of mortars.

Property of the mortar tested	Test	Origin	Standard
Workability	Consistency by flow table	European Standard	EN 1015-3 [92] EN 459-2 [93] EN 413-2 [94]
		British Standard	BS 4551-1 [95]
		American Standard	ASTM C230/ C230M-14 [96] ASTM C1437-15 [97]
	Consistency by penetration	European Standard	EN 1015-4 [98] EN 459-2 [93] EN 413-2 [94]
		British Standard	BS 4551-1 [95]
		Deutsch Standard	DIN 4211:1995-03 [99]
		American Standard	ASTM C187-16 [100]
Water permeability	Water absorption by capillarity test	European Standard	EN 1015-18 [101]
		American Standard	ASTM C1403-15 [102]
		Italian Standard	UNI 10859 [103]
	Water under pressure test	European Standard	EN 1015-21 [90]
Water vapor permeability	Unidirectional drying	European Standard	EN 16322 [104]
	Water vapor permeability	European Standard	EN 1015-19 [105] ASTM E96 / E96M-16 [106]
Adhesion to substrate	Adherence strength	European Standard	EN 1015-12 [107]
Cracking	Modulus of elasticity by frequency of resonance test	European Standard	EN 14146 [108]
	Modulus of elasticity by ultrasound pulse velocity test	European Standard	EN 12504-4 [109]
	Flexural strength test	European Standard	EN 1015-11 [110]
		American Standard	ASTM C348-18 [111]
	Plastic shrinkage test	American Standard	ASTM C1579-13 [112]
	Drying shrinkage test	American Standard	ASTM C596-18 [113]
		Italian Standard	UNI 6555 [114]
	Free shrinkage test	European Standard	Cahier 2669-4 [115]
	Restrained shrinkage test	American Standard	ASTM C1581 / C1581M-18a [116]

(Continued)

Property of the mortar tested	Test	Origin	Standard
Impact resistance	Impact resistance test	European Standard	EN 477 [117]
Compatibility with substrate	Artificial ageing test	European Standard	EN 1015-21 [90]
Durability	Artificial ageing test	European Standard	EN 1015-21 [90]
	Fire reaction test	European Standard	EN 13501-1 [118] EN ISO 1182 [119] EN ISO 1716 [120] EN 13823 [121] EN 11925-2 [122]
	Susceptibility to biocolonization test	American Standard	ASTM D5590-00 [91]

Wastes

3

3.1 Introduction

A waste can be defined as any material, substance, or object that its holder discards, intends to discard, or is required to be discarded [123]. The definition of waste is comprehensive and englobes several types of materials, substances, or objects. Independently of the type of waste, their reduction is a current issue for the environment and the population.

Prevention is the first action in waste's management and can be responsible for a lower volume of wastes generated [123]. Prevention starts in the industry and by cultural awareness. The design of more efficient products and packaging materials, as well as more efficient manufacturing processes, should be improved to prevent waste generation. When prevention is no longer possible, the wastes should be reused in the same economic activity or in others, promoting a circular economy. A waste from one industry can be a raw material of another, avoiding using virgin raw materials and giving a new use to the wastes, extending their lifecycle. The wastes should only be recycled if it is not possible to reuse them [123]. Recycling comes after reuse as it usually involves some energy consumption and it may generate other wastes. If reuse and recycling are not viable solutions, other recovery actions should be considered [123]. Energy valorization is one of the recoveries available. Disposal is the last, and least desirable, option in waste management. A large part of the wastes takes thousands of years to degrade. Thus, with the increase of waste production, landfills are starting to be saturated. In addition to space scarcity, waste disposal in landfill causes environmental impacts such as soil, air, and water pollution [6].

The encapsulation of wastes in hydraulic materials as mortars or concrete is a possible solution for waste management. This incorporation prevents landfill deposition and extends the life cycle of the waste, giving it a new use. Furthermore, wastes can also replace an initial constituent (as the natural aggregate or binder) reducing the global environmental impact of the new material, when compared with the original formulation.

3.2 Origin

There are many types of wastes, all over the world. The wastes can be divided in terms of their nature or origin. In this book the wastes are classified as agricultural, urban, industrial, and hospital wastes.

3.2.1 *Agricultural wastes*

Agricultural wastes are those generated in the agricultural, animal husbandry, or fishing activities. These wastes can be similar to urban wastes (as paper or plastics) or can be specific of these activities.

Eco-efficient Rendering Mortars. DOI: 10.1016/C2018-0-04147-X

Wastes from animal-based and agricultural and fishing activities can be encapsulated in cement-based mortars, concrete, or other cement composites, avoiding landfilling, extending their lifecycle and simultaneously not reducing or if possible even improving some characteristics of the hydraulic elements. In Table 3.1 some examples of the incorporation of these wastes are presented.

Concerning wastes that come from animal sources, the incorporation of chicken feathers, wool, or pig hair has already been tested in mortars as fibers, reinforcing the mortars namely in terms of microcracking behavior. The incorporation of eggshell or mussel shell powders can also be foreseen in mortars or concrete, according to the scientific community. Concerning agricultural wastes, the incorporation of some of these in mortars or concrete, namely natural fibers, has been investigated. The incorporation of fibers in mortars or concrete is beneficial to avoid and prevent cracking propagation and therefore has been investigated by the scientific community (Table 3.1). Natural fibers from agave, banana, coir, palm, date-palm, jute, sisal, rice straw, or kenaf are examples of agricultural wastes whose incorporation in mortars or concrete is already being tested by some authors. The fishing activities are responsible for the production of several wastes, namely fishing nets. Some of them are discarded in the oceans and are harmful to the aquatic environment and animal inhabitant species. The recycling of these nets in mortars has been researched [73] [74].

Table 3.1 Animal-based, agricultural, and fishing activities wastes.

Waste	Composite type	References
Chicken feathers	Mortar	[124]
Wool		[125] [126]
Pig hair		[127]
Eggshell		[128] [129]
Mussel shell	Concrete	[130]
Agave	Mortar	[131]
Date-palm		[69]
Coconut fibre		[70]
Coir		[68]
Jute		[68]
Sisal		[71] [72] [68]
Hibiscus cannebinus		[68]
Cork		[132]
Kenaf		[72]
Fishing nets		[73] [74]
Sisal	Concrete	[133]
Cork		[132]
Rice straw	Cement-based composites	[134]
Bamboo		[134]
Sisal		[135]
Banana		[135]
Eucalyptus grandis pulp		[135]

Table 3.2 Urban wastes' recycling solutions.

Waste	Composite type	References
Plastics	Mortar	[136]–[140]
Glass		[141]–[143]
Broom fibres		[144]–[146]
Rubbers		[147]
Shredded tires		[148]
Electrical cable wastes		[147]
Plastics	Concrete	[136] [149] [150]
Rubbers		[151]
Tires		[152]
Steel fibres from tires		[152]
Glass	Self-compacting concrete	[153]

3.2.2 Urban waste

Urban wastes are those generated by the population in their daily life. The wastes can be more generalist as plastics, packaging materials, paper, glass, alimentary oils, but others more specific are included, such as piles, batteries, electrical cables, or tires.

Table 3.2 presents some examples of recycling solutions that have been studied over the last years with the incorporation of urban wastes in cement-based mortars or concrete. Plastics, glass, rubbers, or tires are some of the most researched wastes. However, less common wastes, such as broom fibers or electrical cable wastes, have also been investigated by the scientific community.

3.2.3 Industrial waste

Industrial wastes are those produced by the industries. In industrial wastes, there are also materials comparable to urban wastes, namely plastics, paper, and glass. The industrial wastes are specific of each industry. There is a wide variety of industrial wastes. Ceramics (white or red), concrete, construction and demolition waste, stone pieces or powders, metallics, polymers, zeolites, textiles, cork are examples of industrial wastes. There are also industrial wastes from the production of energy by the thermal valorization of wastes, namely biomasses, which become ashes, such as bottom and fly ashes.

In Table 3.3, some examples about the incorporation of industrial wastes in cement mortars or concrete are presented.

3.3 Treatment

The wastes, as already mentioned, are materials discarded or that the holder has the intention of discarding. The characteristics of the wastes, at the moment that they are discarded, sometimes are not compatible with the characteristics required to be incorporated in mortars. In Tables 3.4–3.7 some treatments applied to the wastes by several

Table 3.3 Industrial wastes' recycling solutions.

Waste	Composite type	references
Sanitary ware	Mortar	[1] [7] [18] [29] [40] [47] [48]
Red ceramic		[49] [50]
Concrete		[8] [9] [51]
Construction and demolition waste		[10] [11] [154] [155]
Stone powder		[156]
Metallic waste		[157]
Carbon Fibres		[158]
FCC (Fluid Catalytic Cracking)		[159]–[169]
EPS (Expanded Polystyrene)		[170] [171]
GFRP (Glass Fibre Reinforced Polymer)		[65]–[67] [172] [174]
CFRP (Carbon Fibre Reinforced Polymer)		[158]
PE (Polyethylene Polymer)		[52]
Textile fibre		[175] [178]
Glass fibre		[175]
Acrylic fibre		[175]
Forest wood ash		[55] [179] [180]
Wood pellets ash		[53]
Wood ash		[53] [54]
Cacao hush ash		[53]
Molasses ash		[53]
Eucalyptus bark ash		[56] [57]
Palm oil ash		[58] [60] [61]
Olive tree ash		[62]–[64]
Rice hush ash		[61]
Cork ash		[181]
Paper industry ash		[55] [182]
Sanitary ware	Concrete	[12]–[17] [19]–[28] [30] [31]
Tiles		[32] [35]
Red ceramic		[16] [17] [32]–[35]
Concrete		[27] [36]–[39] [41] [42]
Construction and demolition waste		[43]–[46]
Stone slurry		[183] [184]
FCC		[169], [185]–[188]
GFRP		[189]–[194]
CFRP		[195]
Timber ash		[196]
Chestnut wood ash		[197]
Poplar wood ash		[197]
Lechuguilla ash		[198]
Flax ash		[198]
Eucalyptus bark ash		[199]
Pine bark ash		[199]
Paper sludge ash		[53]

(Continued)

Waste	Composite type	references
Kraft pulp fibre	Cement board	[200]
Sanitary ware	Self-compacting concrete	[201]
GFRP		[66] [202] [203]
Olive tree ash		[204]
Rice hush ash		[205]
FCC	Others	[206]–[209]

Table 3.4 Examples of treatments to ceramic, concrete, and stone wastes.

Waste	Type of treatment	Specific treatment	Reference
Red ceramics	Mechanical	Crushed by a jaw crusher and by a roll mill, then sieved	[16] [17]
		Crushed into a jaw crusher and sieved	[33]
		Crushed using a disc-type grinder and sieved	[34]
		Crushed and sieved	[35]
Tiles	Mechanical	Crushed by a ball mill and sieved	[32]
		Crushed and sieved	[35]
Sanitary ware	Mechanical, physical	Crushed by a jaw crusher and by a roll mill, then sieved	[7] [1]–[16] [18]
		Manual broken with a hummer, crushed, gridded, and sieved	[29]
		Manual broken with a hummer, crushed (with a jaw crusher and a cone crusher), grounded, and sieved	[40] [47] [48]
		Crushed in a jaw crusher and sieved	[12] [13] [20] [23]–[27]
		Crushed first with a hammer, then crushed with a jaw crusher and at last sieved	[28]
		Crushed and sieved	[31]
Concrete	Mechanical	Crushed and sieved	[41] [42]
		Crushed in a jaw crusher and sieved	[8] [36]
		Crushed in an impact crusher and sieved	[9]

(Continued)

Table 3.4 *Continued*

Waste	Type of treatment	Specific treatment	Reference
		Crushed first using a jaw crusher and then using a hammer mill sieved at last were sieved	[37] [38] [39]
Construction and demolition waste	Mechanical, physical	Crushing, sieving	[154]
		Crushed in an impact crusher and sieved	[11]
		Crushed (using hammer mill or impact crushing), sieved and magnetic removal of metals	[44]
Stone powder	Mechanical, physical	Dried, crushed in an abrasion machine and sieved	[156]

Table 3.5 Examples of treatments to ashes wastes.

Waste	Type of treatment	Specific treatment	Reference
Forest biomass bottom ashes	Mechanical	Sieving	[180]
Sawdust	Mechanical	Sieving	[196]
Olive tree biomass ashes	Mechanical	Crushing	[63]
Olive tree biomass ashes	Mechanical	Sieving	[63]
Forest biomass bottom ashes	Physical	Washing	[180]
Olive tree biomass ashes	Physical	Combusting	[63]
Forest wood biomass ashes	Mechanical, physical	Dried at 60 °C and sieved	[55] [182]
		Dried, sieved, and the particles under 4 mm were ground in a ball mill for 24 h	[179]
		Sieved	[180]
		Washed and sieved	[180]
		Grounded in a planetary mill for 10 min	[54]
Paper industry pulp biomass fly ash	Mechanical, physical	Dried at 60 °C and sieved	[55] [182]

(Continued)

Waste	Type of treatment	Specific treatment	Reference
Palm oil fuel ash	Mechanical, physical	Dried at 105 °C for 24 h, sieved (only particles below 300 μm were used) and then the particles were grounded	[58]
		Burned at a temperature of 800 °C to 1000 °C. Before used in mortars it was dried at a temperature of 110 °C and sieved	[60]
		Grounded in a ball mill	[61]
Rice husk-bark ash	Mechanical	Grounded in a ball mill	[61]
Olive tree ash	Mechanical, physical	Crushed	[63]
		Combusted	[63]
		Sieved and eliminated the lightweight particles	[63]
		Combusted, sieved, and eliminated the lightweight particles	[63]
		Crushed and combusted	[63]
		Crushed, sieved, and eliminated the lightweight particles	[63]
		Combusted, crushed, sieved, and eliminated the lightweight particles	[63]
Cork ash	Mechanical, physical	Dried at 105 °C for 24 h	[181]
		Dried at 105 °C for 24 h and grounded in a planetary mill	[181]

Table 3.6 Examples of treatments to agricultural, animal-based, and fishing activities wastes.

Waste	Type of treatment	Specific treatment	Reference
Cork	Physical and mechanical	Soak in water, drain by gravity, and with a mechanical press eliminate the water and tannins	[132]
		Heat at 50 °C, soak in water, drain by gravity, and with a mechanical press eliminate the water and tannins	[132]
		Heat at 100 °C, soak in water, drain by gravity, and with a mechanical press eliminate the water and tannins	[132]

(Continued)

Table 3.6 *Continued*

Waste	Type of treatment	Specific treatment	Reference
Banana	Chemical, physical, and mechanical	Kraft pulped, sieved, slurry vacuum dewatered, pressed, crumbed, and stored	[135]
Coir	Mechanical	Cut	[68]
Jute	Mechanical	Cut	[68]
Sisal	Mechanical	Cut	[68]
Hibiscus cannebinus	Mechanical	Cut	[68]
Date-palm fibres	Physical	Washed and dried	[69]
Coconut fibres	Mechanical, physical	Washed, boiled for 2 h, dried at 100 °C in an oven for 24 h, and then cut	[59]
Palm oil fibres	Mechanical, physical, chemical	Sun dried for 6 h, fibres were cut mechanically, washed with tap water and an alkaline treatment using 2% of NaOH solution was performed	[60]
Rice cellulosic fibres	Mechanical	Slurry vacuum dewatering technique	[134]
Bamboo cellulosic fibres	Mechanical	Slurry vacuum dewatering technique	[134]
Sisal	Chemical, physical, and mechanical	Kraft pulped, sieved, slurry vacuum dewatered, pressed, crumbed, and stored	[135]
Eucalyptus grandis pulp	Chemical, physical, and mechanical	Kraft pulped, sieved, slurry vacuum dewatered, pressed, crumbed, and stored	[135]
Eggshell powder	Mechanical	Milled in a ball mill	[128] [129]
	Physical	Heated at 20, 150, 300, and 450 °C	[128]
	Physical	Washed and dried	[128] [129]
Mussel shell	Mechanical, physical	Heated at 135 °C, crushed and sieved	[130]
Chicken feathers	Mechanical, physical	Washed, disinfected, and dried, mill at a roll mill (4 h)	[124]
Pig hair fibres	Physical	Washed and dried	[127]
Wool fibres	Chemical	Plasma treatment	[125]
Fishing nets	Physical and mechanical	Washed and cut	[74]
	Mechanical, chemical	Cut and alkaline treatment	[73]
Fishing nets	Mechanical	Cutting	[73] [74]

Table 3.7 Examples of treatments to industrial and urban wastes.

Waste	Type of treatment	Specific treatment	Reference
PET flakes	Physical and mechanical	Prewashing, gridding, hot washing, drying	[137]
PET pellets	Physical and mechanical	Prewashing, gridding, hot washing, drying, vacuum reactor, extrusion, cooling	[137]
Broom fibres	Mechanical, physical	Cut and washed	[144] [145] [146]
Glass fines or powder	Mechanical, physical	Jaw crusher (60 min), rotation ball mill, sieved	[141]
		Crushed, sieved	[81] [142] [143]
Metallic waste	Mechanical	Mechanically cut	[157]
FCC	Mechanical	Crushed using a ball mill	[168]
EPS	Mechanical	Grounded	[170]
GFRP fibres	Mechanical	Thermolysis at 550 °C, grounded in a ball mill and sieved	[173]
GFRP	Mechanical	Cut, crushed in a shredder and hammer mill, and sieved	[174]
Glass fibres obtained from polymeric composites	Physical	Combusted	[175]
Textile fibres	Mechanical	Cut	[175] [176]
Acrylic fibres	Mechanical	Cut	[175]

researchers are presented. Some physical, mechanical, and chemical treatments can be performed to the wastes (as particles) before their incorporation in mortars. In the following sections these treatments are discussed. At last, the influence of these treatments in the final characteristics of the wastes is also discussed.

The modification of the waste through physical, mechanical, and chemical treatments is performed to improve the quality of wastes used as an aggregate or potential binder in mortars or concrete.

3.3.1 Physical treatments

There are some physical treatments that can or must be performed on the wastes, before their incorporation in mortars. The treatments should be specific for each type of waste. These treatments, as all the other presented in this book, should be avoided whenever possible to reduce the environmental impacts and costs of the new mortars.

Washing and incineration are examples of physical treatments that can be performed on the wastes before their incorporation in mortars. A washing stage can be adopted to remove impurities or to disinfect the wastes [130] [136]. This process should be performed whenever the waste presents some substance that can be released in contact with water and can contaminate the mortars cement-matrix or the environment. Fishing nets wastes were in contact with saltwater during their life cycle. These salts can deteriorate the mortars. The crystallization of salts in the mortars internal pores can result in internal microcracks, which will influence the hygric and the mechanical behavior of the mortars. To avoid the anomalies that result from the salts, these fibers should be washed before their incorporation. There are other wastes that can contaminate the environment, namely due to the release of heavy metals and these also should be submitted to a washing treatment. Some authors also wash their wastes to eliminate the thinnest fraction of the waste. Incineration is another treatment that can be performed on the wastes before their incorporation in mortars. The incineration eliminates most of the organic matter present in the waste. If the intention is to use only the inorganic part of the waste, the incineration can be a possible treatment. In some cases, incineration takes place to solve other issues associated with the wastes. The wastes occupy space, landfills start to be overwhelmed and incineration comes as a solution to reduce the mass and volume of the wastes, reducing the space required if they are sent to landfill. Municipal solid wastes are examples of wastes that are submitted to an incineration process to reduce their volume and mass [210]. The volume of these wastes is reduced by about 90% [210] when submitted to high temperatures.

3.3.2 Mechanical treatments

Mechanical treatments are those that require an equipment to perform the treatment. Crushing, milling, and sieving are the treatments most performed on the wastes. In Tables 3.4–3.7, it is possible to verify that many of the wastes pass through at least one of these processes. That is because wastes are discarded in the size they have in that specific moment. In most of the cases, the wastes do not present the size characteristics required to replace an initial constituent of the mortars and have to be submitted to a treatment process prior to their application in mortars. The treatments should be reduced to the minimum to reduce the environmental impacts and costs of the process.

Crushing and milling processes are sometimes necessary to reduce the size of the waste particles until a size that is compatible with that of the aggregate and the binder particles. Sometimes milling can also be performed to improve the reactivity of the wastes [58] [159], namely the ones that intend to replace the cement. Thinner particles have more ability to react.

Crushing is an action performed sometimes before milling, whenever the particles are very large. The crushing process can be manual or mechanical. Milling is a mechanical process that can reduce the size of the particles to several micrometers. There are several types of mills: jaw, ball, roll, knife, hammer, and others. The mills function and act differently but the intent of all types is to reduce the size of the particles. For example, concerning the ball mill the wastes are put together with spherical balls, usually in metal, inside a mill that slowly rotates along a horizontal axis. With the rotation, the balls fall on the waste pieces, breaking them. Usually the time that the pieces stay in the mill is fixed. On the other hand, the roll mill is composed by two cylinders (usually made of metal) that rotate along a horizontal axis with the same velocity but in opposite directions. The particles are introduced between the cylinders and the particles are crushed by cutting. The distance between the mill rolls is reduced in each single cycle to gradually reduce the size of the particles. The process ends when the specific size is achieved. In Fig. 3.1, a roll mill is represented. The millers used depend on their availability but also on the type of waste intended to mill. According to Saikia and Brito [136], propeller crushers or blade mills are used to grind the plastic waste.

Sieving is another mechanical treatment usually performed. This treatment can be applied in several situations, for example when the user only uses part of the size

Fig. 3.1 Example of a roll mill used to reduce the size grain of waste particles.

distribution of the waste or, on the contrary, wishes to eliminate a part of the size distribution of the waste. Sieving can also be performed to the waste to guarantee that the grain size of the waste particles is similar to the ones of the original constituent that has to be replaced by the waste.

3.3.3 Chemical treatments

Chemical treatments can be performed on the wastes to improve their characteristics, namely reducing their porosity. Carbonation, impregnation, or microbial-induced carbonate precipitation are examples of treatments used to improve the waste characteristics. These treatments, similar to the ones presented previously, should be carefully chosen as they have associated environmental impacts, thus these impacts should be evaluated beforehand.

Carbonation is a process usually performed to wastes that intend to replace the natural aggregates. This process is responsible for a pore refinement of the cement paste matrix [211], being responsible for a consolidation of the matrix. There are several wastes that can be submitted to this treatment, such as construction and demolition wastes. These wastes become less absorbent, denser, and stronger [212] with this type of treatment. Another treatment that can be used in waste aggregates is their impregnation. This process consists in injecting a compact material, as a polymer [213], in the pores of the waste aggregates, filling the pores and consolidating the wastes. The use of microbial-induced carbonate precipitation technique is also a chemical treatment that can be used for several waste aggregates [214]. This procedure uses bacteria that promote the precipitation of calcium carbonate [214] and consequently improves the compactness of the aggregates. A plasma treatment was also used by some authors to modify the surface of the fibrous materials [125]. Some wastes can also be subjected to alkaline treatments. Spadea et al. [73], for example, incorporated fishing nets in cement-based mortars by treating these nets in a solution of 10.4 g sodium hydroxide and 999 mL of distilled water for 120 h at a temperature of 60 ± 2 °C for an alkaline treatment of the fibers. Also Pelisser et al. [151] washed rubber wastes with sodium hydroxide to increase the hydrophilicity of the rubber particles' surface.

3.3.4 The influence of waste treatment on its final characteristics

As described, the wastes can be submitted to several treatments before their incorporation in mortars. Examples of mechanical, physical, and chemical treatments were presented previously. These treatments can be responsible for some modifications of the waste particles, such as shape, roughness, size distribution, porosity, or for some change in the chemical characteristics of the particles.

3.4 Properties

Following some specific physical, mechanical, and chemical characteristics of the wastes are presented. The physical properties are the most influenced by the waste's treatment, especially the ones that involve the reduction of the wastes size (as crushing or milling treatments). The mechanical characteristics of the wastes are less dependent on the treatments; however, in some cases, namely some chemical treatments (as carbonation, impregnation, or microbial-induced carbonate precipitation technique), the treatment can be responsible for a reduction of the porosity, which improves the mechanical behavior of the wastes. The chemical properties of the wastes are related to the nature of the wastes' particles. Some treatments could also influence the wastes' chemical composition, for example in the case of wastes calcination.

3.4.1 *Physical properties*

Shape, roughness, porosity, water absorption, size and size distribution are physical properties of the wastes that influence the interface between these waste materials and the cementitious matrix of the mortars. Next, it is described how do these properties influence the mortars behavior.

3.4.1.1 Shape and roughness

As mentioned, the treatments that involve size reduction have a great influence in the shape and roughness of the particles. The crushing and milling treatment processes, in general, result in more angular and rough particles, when compared with the natural aggregate. Natural aggregate is mostly spherical and presents a smooth surface [138]. These characteristics are related to the movements of the waves or rivers that thin out the particles edges. The wastes that were artificially resized present more edges when compared with natural particles. This has been reported by several authors [23] [34] [65] [171]. These angularities influence the shape of the aggregates as well as increase their surface contact area. This high surface area influences the rearrangement of the particles, which is worse than that of spherical particles. An angular shape increases the volume of empty voids when compared with more spherical particles. A higher surface area can also promote an increase of kneading water needs of the mortars. The water required to lubricate the angular particles is higher than the one necessary for spherical particles. On the other hand, particles with high surface area are characterized by a higher area of contact between the aggregates and the binder matrix, which can promote a better mechanical bond between these two elements.

The surface of the particles also influences their adherence to the cementitious paste [138]. Particles with smooth surfaces present a lower contact area than particles with a rough surface. The particles with rough surface are characterized by more edges, cavities, or irregularities that promote the adherence between the wastes and the matrix of the mortars.

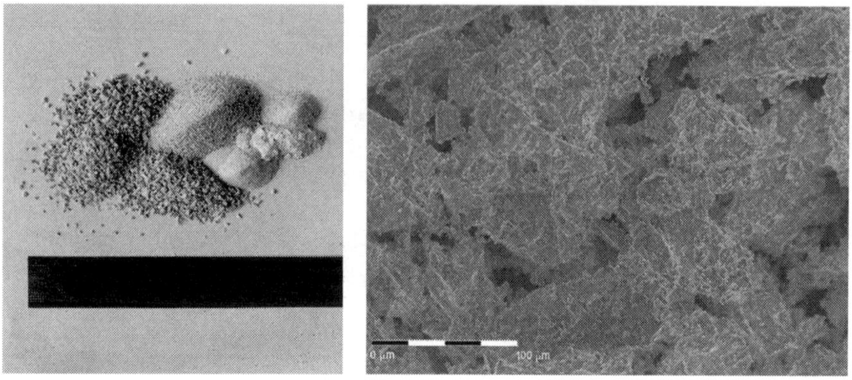

Fig. 3.2 Sanitary ware waste.

In Figs. 3.2–3.5, some examples of wastes from different industries are presented. In Fig. 3.2, particles of sanitary ware waste that went through manual crushing and afterwards were milled in jaw and roll mills [7] are presented. The resulting sanitary ware was characterized by particles with an irregular shape and rough surface [175]. In Fig. 3.3, a waste from the polymer industry that was not submitted to any treatment process before its application in rendering mortars is presented. Glass fiber–reinforced polymer powder is a waste from the industry that was produced during the cutting of polymer elements and afterwards was aspirated by the central suction system [172]. The waste also presents an irregular shape, but it does not seem to present rough particles [172]. Forest biomass bottom ashes' waste is represented in Fig. 3.4. The ashes were incorporated in mortars without any treatment [215]. Similar to the previous wastes the biomass presents irregular particles that are characterized by a rough surface [215]. In Fig. 3.5, a fluid catalytic cracking waste from the petrochemical industry is represented. This waste also was not submitted to any treatment. It presents a spherical shape and a smooth surface [175], contrary to the other wastes presented before.

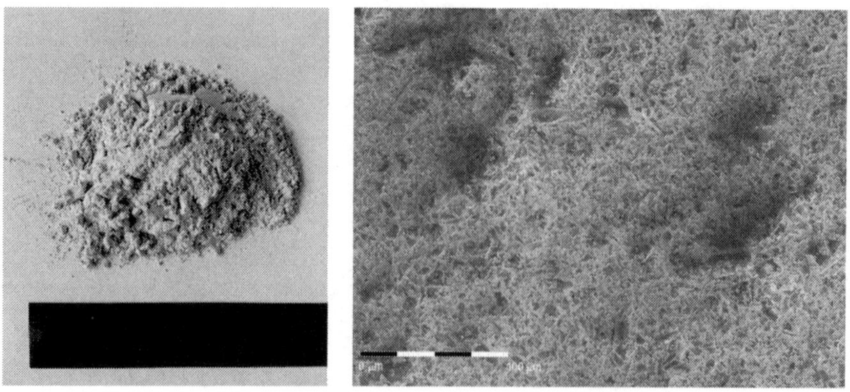

Fig. 3.3 Glass fiber–reinforced polymer (powder waste).

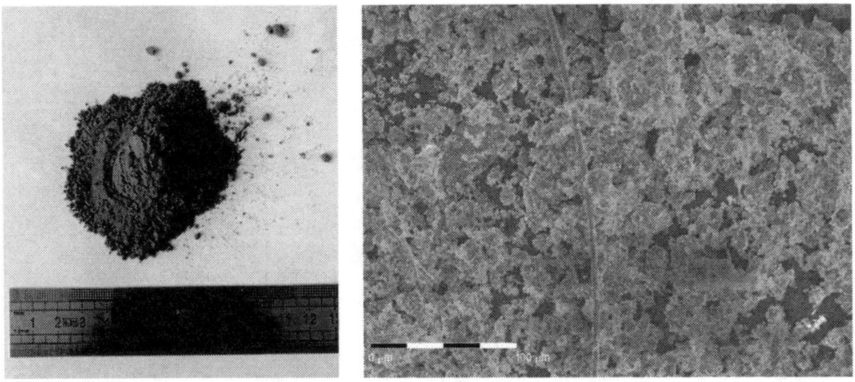

Fig. 3.4 Forest biomass ashes waste.

Table 3.8 presents some results of specific surface areas of some wastes incorporated in mortars or concrete as aggregates or binders. The natural aggregate presents a surface area between 0.09 and 2.70 m^2/g and cement between 0.35 and 1.81 m^2/g. Some wastes present a high specific surface area, such as forest biomass ashes, which in some cases achieve 40 m^2/g. The specific surface area of the biomass ashes presents a high range of 9–40 m^2/g. This variation of values is related to several aspects such as incineration process, temperature achieved during burning, or biomass fuel used. There are several factors that influence the ashes surface area. Fluid catalytic cracking is one of the wastes that present higher specific area, which is related to the cavities of the zeolite materials. These materials are characterized by internal cavities/tunnels responsible for incrementing the specific surface area of the wastes, around 90 and 120 m^2/g of the product.

Fig. 3.5 Fluid catalytic cracking waste.

Table 3.8 Specific surface area: results from several authors.

Particles	Specific surface area (m^2/g)	References
Sand	009–2.70	[75] [175]
Cement	035–1.81	[61] [66] [75] [140] [216]
Eggshell powder	0.31–0.37	[128] [129]
Sanitary ware	0.534–1.35	[27] [31] [175]
Red ceramics	0.555	[34]
Construction and demolition waste	6.58	[27]
Tiles	0.5582	[32]
Stone powder	0.245	[156]
Stone slurry	0.7128	[183] [184]
Forest biomass ashes	8.63–40.00	[55] [175]
Paper biomass ashes	8.00	[55]
Rice husk-bark ash	0.95	[61]
Palm oil fuel ash	0.72	[61]
Eucalyptus wood ash	3.25	[56]
Palm oil fuel ash	0.72–5.1	[58] [61]
Rice husk-bark ash	0.95	[61]
Electric furnace dust	3.7–4.6	[216]
Fluid catalytic cracking	93.62–118.00	[160] [165] [168] [169] [175] [186]
GFRP powder	4.12	[175]

3.4.1.2 Size and size distribution

The size and the size distribution of the particles influence their application field and the performance of the particles whenever incorporated in mortars. The waste particles that intend to replace the natural aggregate in mortars should present a maximum size of 5 mm. These particles should also present a well-graded size distribution curve, meaning that they should have particles of all sizes to achieve a good rearrangement. A proper rearrangement of the particles will promote high compacity of the mortars, decreasing the volume of air voids. In Fig. 3.6, an example of a well-graded curve is presented. This curve has been used as a reference in several scientific studies about the incorporation of wastes in rendering mortars.

The particles that intend to replace the binder should present a size lower than 63 μm. Nevertheless, in those cases, the thinner the particles the more reactive they are.

3.4.1.3 Porosity

The porosity of the waste particles, especially open porosity, influences the adherence between the particles and the matrix of the mortars. The open porosity of the particles

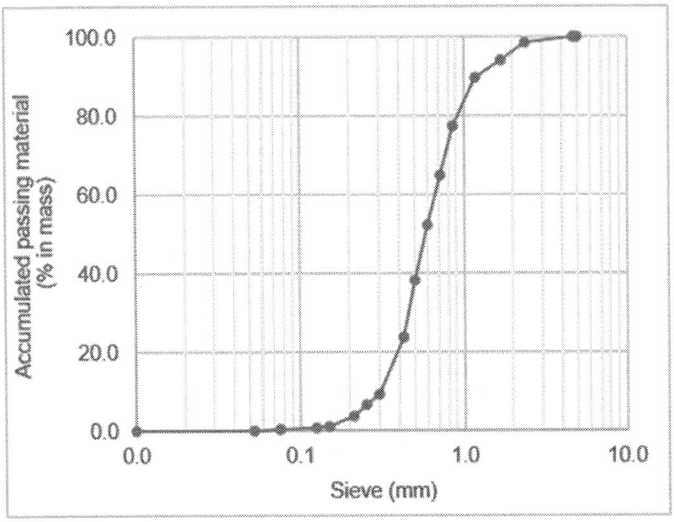

Fig. 3.6 Well-graded size distribution curve for aggregates in rendering mortars (example).

in general influences their ability of water absorption, which promotes a mechanical adherence between the waste particles and the cementitious matrix, as will be presented in the following section. The porosity of the wastes can also promote chemical reactions between these particles and cement.

3.4.1.4 Water absorption

In the fresh mix, the solid constituents absorb part of the fresh paste composed of water and cement. As the aggregates absorb this paste, part of the cement compounds hydrates in the pores of the aggregates creating a stronger adherence between the aggregate and the cement matrix. The higher porosity of the wastes particles is responsible for major water absorption and can lead to a strong bond between the aggregate and the matrix. The water absorption can also promote a delayed hydration of the cement matrix. Wastes in general, due to their high porosity and high specific surface, are more absorbent than natural aggregates. In Table 3.9, some examples of the water absorption of wastes are presented. As can be noticed, the organic wastes (as pig hair, coir, jute, sisal, or hibiscus cannebinus) are highly absorbent, and the percentage of water absorbed is in general higher than 100% of their mass and in some cases achieves 280%. White ceramics (sanitary ware or tiles) absorb an amount of water similar to the natural aggregate. Red ceramics are more porous than white ones, which results in more absorbent materials, absorbing in some cases 12% of water related to the mass of the particle.

3.4.1.5 Density

Density is a property that allows evaluating the compactness of the waste's particles. It can be measured by the ratio between the wastes' mass in a volume unit. Waste

Table 3.9 Water absorption of wastes: results from several authors.

Particles	Water absorption (%)	References
Sand	0.41–3.0	[9] [40] [47] [66] [132] [144] [216]
Pig hair fibre	95	[127]
Mussel shell	2.56–4.12	[130]
Coir	180	[68]
Jute	281	[68]
Sisal	200	[68]
Hibiscus cannebinus	285	[68]
Textile fibres	160.2	[175]
Acrylic fibres	116.9	[175]
Glass fibres	39.4	[175]
Palm oil fibres	0.79	[60]
Sanitary ware	0.3 –1.5	[12] [13] [16] [17] [20] [23]–[26] [40] [47] [175]
Tiles	0.84	[35]
Red ceramic	8.6–12.2	[16] [17] [33] [35]
Concrete	3.9–9.6	[9], [36]–[39], [41], [42]
Construction and demolition waste	2.1–14.29	[11] [43] [44]
Olive tree ashes	18.9–21.8	[63] [64]

density includes only the particles and their specific volume, and bulk density includes the air volume among particles. By combining these two measurements, it is possible to understand the ability of the waste particles to rearrange themselves [175]. The shape of the waste particles directly influences the bulk density. Irregular and angular particles increase the volume of air among particles [175]. Therefore, the treatment process also affects this property. The treatment process that involves a reduction of porosity, such as carbonation or impregnation, affects the waste density and the bulk density of the particles. In Table 3.10, the density of several wastes is presented. Fiber wastes are an example of wastes that present low densities. Wool fibers and coconut fibers, for example, present densities of 1.32 and 1.45 g/cm^3, respectively. Polymers also present low density as PET (Polyethylene terephthalate) flakes or PET pellets that present densities of 1.28 and 1.30 g/cm^3, respectively. On the other side, ceramics, stone, and concrete present densities between 2.1 and 2.9 g/cm^3, values closer to those of the natural aggregates. Some ashes also present a high density, similar to the cement's, for example palm oil fuel ash that in some cases achieves 3.1 g/cm^3.

3.4.2 Mechanical properties

In render applications, the wastes more commonly submitted to mechanical tests are fiber wastes. These fibrous wastes can be submitted to tensile strength tests to identify some of their properties, namely tensile strength, modulus of elasticity, and elonga-

Table 3.10 Density of wastes: results from several authors.

Particles	Density (g/cm^3)	Bulk density (g/cm^3)	References
Sand	**2.5–2.65**	**1.47–1.73**	[9] [51] [66] [137] [139] [140] [142] [144] [175] [216]
Cement	**2.98–3.13**	**0.96–1–01**	[51] [54] [66] [81] [137] [141] [142] [169] [175]
Eggshell powder	2.5	–	[128] [129]
Mussel shell	2.65–2.73	–	[130]
Wool fibres	1.32	–	[125]
Sisal fibres	1.45	–	[72]
Kenaf fibers	1.5	–	[72]
Coconut fibres	1.28	–	[59]
Palm oil fibres	2.13	–	[60]
Fishing nets	1.13	–	[74]
Sanitary ware	2.3–2.969	1.04–1.319	[1] [7] [12] [13] [16] [17] [18] [20] [23] [24] [25] [26] [29] [31] [175]
Tiles	2.362–2.65	–	[32] [35]
Red ceramic	1.948–2.165	1.032	[16] [17] [33] [35]
Concrete	2.2–2.4	0.842–1.393	[9] [36] [37] [38] [39] [41] [51]
Stone slurry	2.72	–	[183] [184]
GFRP powder	1.87	0.59–1.3	[65] [172] [175]
Forest biomass ash	2.455	0.695	[175]
Textile fibres	–	0.010–0.017	[175]
Acrylic fibres	–	0.004–0.005	[175]
Glass fibres	–	0.016–0.020	[175]
Glass powder	–	0.832	[81] [142]
Recycled PET fibres	1.32	–	[74]
PET flakes	1.28	0.47	[137]
PET pellets	1.30	0.73	[137]
PET	–	0.326–0.408	[139]
PET	–	0.547	[140]
Polycarbonate plastic	–	0.612	[140]
Electric furnace dust	3.8	–	[216]
Broom fibres	0.99	–	[144]
Tyre rubbers	1.31	–	[147]
Powder rubber	1.20	–	[147]
Micronized tyre fibres	1.38	–	[147]
Electrical cable wastes	1.33	–	[147]
Steel fibres from tire waste	3.0	–	[152]
Fluid catalytic cracking	2.45–2.71	0.87	[162] [168] [169] [175]
EPS	0.013–0.018	–	[170] [171]
Forest biomass ash	2.65–2.75	–	[54] [180]
Eucalyptus wood ash	2.615–2.619	–	[56] [57]
Palm oil fuel ash	2.22–3.15	–	[58] [61]
Rice husk-bark ash	2.16	–	[61]
Olive tree ash	1.86–2.18	1.37	[63] [64]
Cork ash	2.601	–	[181]

tion. These characteristics are extremely important because fibers, in general, are incorporated in mortars to reinforce them and to reduce their cracking susceptibility. Fibers mechanical properties influence the reinforced render mortars' behavior.

The tensile strength of the fibers should be higher than the mortars' tensile strength. A standard cement-based mortar, with a volumetric ratio of 1:4 (cement:sand) and a cement class 32.5 MPa, presents a flexural strength between 1.0 MPa and 5.0 MPa

Table 3.11 Tensile strength of waste fibers: results from several authors

Waste fibres	Diameter (μm)	Length (mm)	Tensile strength (MPa)	Elongation (%)	Modulus of elasticity (GPa)	References
Wool	19	16	240	–	2	[125]
Wool (treated with plasma)	19	16	290	–	3	[125]
Wool	–	15 and 30	–	–	–	[126]
Pig hair	230	35.7	99.2 ± 41.7	–	–	[127]
Textile fibres	500	15 and 30	34.6 ± 0.8	20.1 ± 0.3	0.261 ± 0.027	[175]
	1000	40	7.44	–	24.40	[176]
Acrylic	10	15 and 30	41.4 ± 1.8	11.6 ± 2.1	5.241 ± 0.209	[175]
Glass	20	15 and 30	14.3 ±2.0	3.0 ± 0.4	10.128 ± 1.546	[175]
Coir	–	20, 30 and 40	50.89	17.6 mm	–	[68]
Jute	–	20, 30 and 40	60.14	13.1 mm	–	[68]
Sisal	–	20, 30 and 40	58.16	6.0 mm	–	[68]
Hibiscus cannebinus	–	20, 30 and 40	76.04	6.7 mm	–	[68]
Sisal	–	25 and 45	670	4	30	[71]
Sisal	120–140	5	530–640	3–7	17–22	[72]
Kenaf	15–30	5	350–600	2.5–3.5	40	[72]
Fishing nets	330	25.4, 31.8 and 38.2	302–348	–	0.728	[73]
Fishing nets (with alkaline treatment)	330	25.4, 31.8 and 38.2	289–342	–	0.724	[73]
Fishing nets	350	20, 30 and 40	440	–	3	[74]
Recycled PET fibres	700	30 and 40	450	–	20	[74]
Rice cellulosic fibres	5.28	0.9	–	–	–	[134]
Bamboo cellulosic fibres	13.13	2.5	–	–	–	[134]
Sisal	13.5	1.65	–	–	–	[135]
Banana	15.3	1.95	–	–	–	[135]
Eucalyptus grandis pulp	10.9	0.66	–	–	–	[135]

(Continued)

Waste fibres	Diameter (μm)	Length (mm)	Tensile strength (MPa)	Elongation (%)	Modulus of elasticity (GPa)	References
Broom fibres	380–510	20 and 50	210–250	55	5	[144] [145] [146]
Steel fibres from tire waste	420 ± 330	37 ± 11.8	870	–	–	[152]
GFRP fibres	19.8 ± 0.4	–	618 ± 79	–	77	[173]
Palm oil fibres	19.4	25	35.33	0.05	–	[60]

[51] [142] [172]. In this sense, the fibers to reinforce the mortars should present strength higher than the mortar's, that is higher than 5 MPa. In Table 3.11, the tensile strength of several organic and inorganic fibers is presented. The tensile strength is quite higher than the cement matrix's and in this sense these fibers are able to reinforce the cement matrix, in terms of tensile strength.

Another important characteristic of the fibers is their modulus of elasticity. The modulus of elasticity of a cement-based mortar, with a volumetric ratio of 1:4 (cement:sand), and with a cement class 32.5 MPa, is usually between 7 GPa and 11 GPa [51] [142] [172]. The fibers should have a modulus of elasticity above that of the mortar to distribute the tensile stresses in the mortar and decrease cracking. If their modulus of elasticity is quite lower than that of the mortar, the fibers are too deformable. Thus, the tensioned fibers will deform without creating a barrier to cracking. Nevertheless, these fibers may still be effective in reducing plastic shrinkage cracking, before the mortar increases its stiffness. In Table 3.11, the modulus of elasticity of several fibers is presented. Some present a high modulus of elasticity, such as sisal fibers, kenaf fibers, or GFRP (glass fiber reinforced polymer) fibers.

Table 3.12 Chemical composition of several types of glass wastes.

Waste	Amber glass cullet	Uncolored glass cullet	Green glass cullet	Pyrolysis glass fibre composite
SiO_2	70.2	71.2	70.4	54.3
CaO	10.3	10.7	10.1	17.2
Al_2O_3	1.9	1.8	1.8	15.2
Fe_2O_3	0.4	-	0.5	8.0
MnO	-	-	-	4.7
MgO	2.6	2.5	2.8	-
Na_2O	13.2	13.4	12.9	-
K_2O	0.6	0.5	0.4	-
SO_3	1.3	-	0.2	-
Cr_2O_3	0.1	-	0.3	-
Reference	[141]	[141]	[141]	[173]

Table 3.13 Chemical composition of ceramic wastes

Waste	Electric isolator (white ceramic)	Sanitary ware		Natural glass tiles	Black glassed tiles	Non-glassed tiles	Construction and demolition waste	Stone slurry (marble and limestone)
SiO_2	70.9	67.91	70.8	63.57	64.04	63.69	48.7	0.91
CaO	0.76	2.41	0.7	1.56	1.64	1.72	17.98	54.29
Al_2O_3	21.1	22.01	19.0	19.50	19.18	19.35	9.83	3.72
Fe_2O_3	0.81	1.41	0.9	7.24	6.96	7.11	2.49	0.40
MnO	-	0.07	-	0.06	0.07	0.07	0.04	-
MgO	0.24	0.29	0.3	1.36	1.37	1.38	1.85	0.30
Na_2O	1.47	1.91	2.7	3.80	3.73	3.77	0.86	-
K_2O	3.57	2.79	4.6	1.26	1.30	1.26	3.17	-
SO_3	-	0.07	-	0.02	0.01	0.02	2.64	0.09
P_2O_5	-	0.17	-	0.14	0.14	0.14	0.12	-
TiO_2	0.33	0.45	0.3	0.88	0.85	0.89	0.27	-
ZnO	-	0.16	-	-	-	-	-	-
Cl^-	-	-	-	-	-	-	-	0.03
Reference	[40]	[27]	[31]	[32]	[32]	[32]	[27]	[183]

Table 3.14 Chemical composition of waste ashes.

Waste	Agave ashes	Rice husk-bark ash	Palm oil fuel ash			Olive biomass ashes
SiO_2	1.3–1.5	84.75	65.30	63.7	61.66	55.35
CaO	64.6–81.2	2.78	6.42	6.0	9.86	16.57
Al_2O_3	–	0.16	2.56	3.7	5.13	5.39
Fe_2O_3	0.2–0.8	0.00	1.98	6.3	5.29	2.25
MgO	0–16.1	2.32	3.08	4.1	4.17	2.64
Na_2O	–	0.37	0.36	–	0.49	–
K_2O	12.5–15.0	2.57	5.72	9.15	8.42	7.67
SO_3	0–0.78	0.60	0.47	1.6	–	0.24
TiO_2	–	–	–	0.30	–	–
P_2O_5	0–3.6	–	–	4.3	3.71	–
Cl	–	–	–	0.50	0.162	–
SrO	0.1–0.2	–	–	–	–	–
Reference	[131]	[61]	[61]	[58]	[60]	[63]

Table 3.15 Chemical composition of wastes ashes (continuation).

Waste	Wood ash	Wood and cacao ashes	Wood pellets ashes	Eucalyptus ashes		Paper sludge ashes	Pulp and paper ashes
SiO_2	73.01	34	24	11.6	36.1	18	28
CaO	2.64	26	27	34.9	28.0	56	25.4
Al_2O_3	11.93	2.8	3.2	4.4	8.4	8.6	6.2
Fe_2O_3	3.38	1.7	2.7	2.6	4.1	1.3	2.2
MgO	1.03	3.3	4.9	–	3.5	2.9	5.0
Na_2O	0.99	0.74	1.3	1.4	0.9	0.77	3.3
K_2O	4.14	7.8	9.35	–	5.4	0.93	3.2
SO_3	< 0.05	3.9	3.2	–	–	1.5	–
TiO_2	0.48	–	–	0.25	0.3	–	0.3
MnO	–	–	–	1.4	0.9	–	0.7
P_2O_5	0.59	3.6	2.3	1.6	3.2	0.32	0.9
Cl^-	0.009	–	–	4.4	–	–	–
SrO	–	–	–	0.21	–	–	–
V_2O_5	–	–	–	0.18	–	–	–
ZnO	–	–	–	0.057	–	–	–
NiO	–	–	–	0.040	–	–	–
Rb_2O	–	–	–	0.037	–	–	–
CuO	–	–	–	0.022	–	–	–
ZrO_2	–	–	–	0.020	–	–	–
Br	–	–	–	0.007	–	–	–
Reference	[54]	[53]	[53]	[56]	[57]	[53]	[182]

3.4.3 Chemical properties

3.4.3.1 Chemical composition

The chemical characteristics of each waste depend on its nature. From Tables 3.12–Table 3.16 the chemical composition of diverse wastes is presented.

3.4.3.2 Pozzolanicity

A pozzolan is a material that is not naturally a binder but contains amorphous phases of silica and/or alumina that, under given circumstances, in the presence of water and at current temperature, can be combined with the calcium oxide present in cement or lime, giving rise to stable elements with hydraulic binder properties [82]. Some

Table 3.16 Chemical composition of zeolites wastes.

Waste	FCC	Ecat	EPcat
SiO_2	48.2	51.69	45.06
CaO	46.0	1.73	2.30
Al_2O_3	0.95	41.12	32.91
Fe_2O_3	-	1.08	3.04
MgO	-	0.41	1.10
Na_2O	0.50	0.42	0.37
K_2O	-	0.06	0.06
SO_3	0.04	0.48	8.09
Reference	[159]	[161]	[161]

Table 3.17 Amount of $Ca(OH)_2$ fixed by wastes.

Waste	Forest biomass ashes	Sanitary ware	FCC
Mg of $Ca(OH)_2$ fixed	232	194	926-1540
Reference	[218]	[218]	[164] [218]

wastes, due to their chemical composition, can act as a pozzolan. Ashes wastes are wastes that can present some pozzolanicity.

The amount of calcium hydroxide fixed on a pozzolanic material is an indicator of its pozzolanicity activity [217]. In this sense, some authors evaluate this fixation. In Table 3.17, some examples of wastes able to fix some amount of $Ca(OH)_2$ are presented.

The concentration of CaO and OH^- ions in a solution with cement can also indicate the pozzolanicity of the wastes, according to EN 196-5 [219]. In Table 3.18 some examples of wastes considered or not pozzolanic according to the European Standard are presented. Rajamma et al. [182] found from this method that forest biomass ashes presented high pozzolanicity contrary to pulp and paper industry ashes that tested negative. Also, Garcia and Sousa-Coutinho [179], found that wood grounded ashes, by EN 196-5 [219], were classified as no pozzolanic. Contrary, Ramos et al. [54] found that wood ashes were tested positive in pozzolanicity by the same standard. In another research, Ramos et al. [181] evaluated the pozzolanicity of cork ash by the same method and noticed that the concentration of CaO was about 13.6 mmol/L and OH^- was about 48 mmol/L, which was lower than the saturation, leading to a negative pozzolanic result.

Pozzolanicity can also be evaluated by the mechanical strength of mortars when the hydraulic binder (cement) is replaced with a pozzolanic material. The decrease

Table 3.18 Pozzolanicity positive or negative by EN 196-5 methodology.

Waste	Forest biomass ashes	Grounded wood ashes	Wood ashes	Chestnut ashes	Poplar ashes	Pulp and paper industry ashes	Cork ashes
Pozzo-lanicity	Positive	Negative	Positive	Nega-tive	Nega-tive	Nega-tive	Nega-tive
Refer-ence	[182]	[179]	[54]	[197]	[197]	[182]	[181]

Table 3.19 Activity index of some wastes.

Waste	Coal fly ash	Forest bio-mass ashes	Biomass fly ash	Sanitary ware	FCC
Activity index at 28 days	76.0	72.1	76.1	60.3	43.4
Activity index at 90 days	87.9	76.7	75.0	64.4	50.6
Reference	[57]	[218]	[57]	[218]	[218]

of strength is low when a waste is pozzolanic, because the waste, after combination with the calcium hydroxide, acts as a hydraulic binder. Some authors also evaluate the pozzolanicity by this method. According to the European Standard 196-1 [220], a material is considered pozzolanic if the mortar with its incorporation in substitution of cement presents a compressive strength at 28 days at least 75% of the strength of the reference mortar and at 90 days at least 85%. In Table 3.19, the ratio between the compressive strength of the blended mortar and the reference mortar (designated by activity index) for some wastes is summed up.

3.5 Chapter conclusions

Wastes can come from several origins and can be subjected to a great variety of treatments before their application in new materials.

The treatments should only be applied whenever they are essential, as they are responsible for several environmental impacts, not only in energy spending but also in other environmental issues such as air and soil pollution. To be incorporated in mortars, wastes can be submitted to physical, mechanical, or chemicals treatments that will modify the waste's original properties. The properties of the wastes are influenced by the wastes' treatment and the nature of each waste.

To predict the performance of the wastes, when incorporated in mortars, the knowledge of the wastes' treatments and properties is essential.

Mortars with incorporation of wastes: characteristics of fresh mortars

4

Fresh properties are extremely important in render mortars. This specific type of mortars should be workable to be applied in vertical façades (Fig. 4.1). The fresh characteristics of mortars are influenced by the incorporation of wastes as will be seen in this chapter. The properties of the wastes, such as their porosity, roughness, or shape, influence the rearrangement of the particles and their water absorption; these characteristics of the particles influence the workability of the mortars, their bulk density, and their ability to retain water. Concerning the influence of the wastes' incorporation on the workability of the mortars, their water/binder ratio, bulk density, and water retention are discussed. Some examples of wastes incorporation are also presented.

4.1 Workability

Workability is one of the most important characteristics of a mortar used for rendering. Without proper workability, a mortar cannot be applied on vertical surfaces. Thus, the workability of the mortars is usually the first property to be defined. The water/binder ratio of a mix is defined taking in account the purpose of achieving that workability. The amount of water required in a mix is greater than the water required for cement hydration, as part of the water is necessary to lubricate the surface of the solid constituents of a mortar. This lubrication is necessary for an adequate workability. The amount of water required depends on the aggregate's size distribution curve, porosity, water absorption, shape, and porous surface [138] [140]. The incorporation of wastes in mortars modifies these properties, which has a direct influence on the workability of the mortars.

In Fig. 4.2, the relative consistency of modified mortars (with incorporation of waste materials) and a reference mortar (without wastes) obtained in several studies is presented. In the figure, the wastes incorporated as aggregates, such as fillers, binders, or fibers, are separately identified. The studies were differentiated because each type of waste interferes differently with each property of the mortars.

Filler wastes, composed by thin particles less than 150 μm, are those that can improve the workability of the mortars. In Fig. 4.2, some filler wastes that improved workability are represented, such as glass powder, concrete powder, red ceramic powder, or sanitary ware powder. Fillers fill the air voids between the aggregates making the paste more fluid and more plastic.

When the wastes are incorporated as aggregates, replacing natural aggregate, in the same or in a similar size distribution, the workability is, in general, negatively affected. This decrease of workability is due to the wastes' properties that are

Eco-efficient Rendering Mortars. DOI: 10.1016/C2018-0-04147-X

Fig. 4.1 Render mortar with wastes incorporated (application).

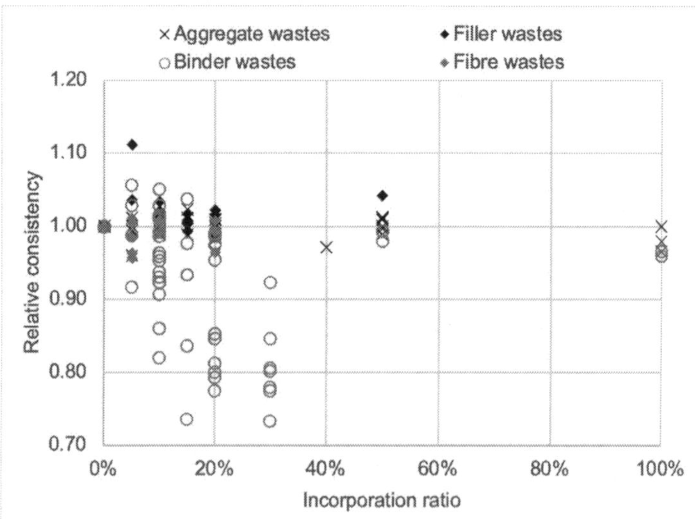

Fig. 4.2 Workability of mortars with aggregate, filler, binder, or fiber wastes [7]–[9] [18] [49] [51] [54] [55] [64] [67] [126] [137] [142] [143] [148] [154] [155] [162] [168] [169] [172] [179]–[182] [215] [218].

influenced by their origin and treatment, as discussed in the previous chapter. The wastes are, in general, more porous and rougher than natural aggregates and these properties directly influence the workability of the mortars. The incorporation of these wastes can increase their consistency and decrease their workability. However, the studies analyzed show that the decrease of workability, in most of the cases, is slight (below 10%).

Fiber wastes in general decrease the workability of the mortars [59] [74] [176] [190]. The water absorption and adsorption of the fibers are the main reason for an increase in the consistency of the mortars. Fibers retain a great amount of water. Part of this water is absorbed, and another part is retained on the surface of the fibers. It is necessary to increase the water/binder ratio or to add an admixture to achieve an adequate workability.

The wastes incorporated in mortars aiming at replacing the cement are designated as "binder wastes". These wastes are the ones that most influence the mortars' workability (e.g., Fig. 4.2). They can be responsible for a reduction of workability that can achieve almost 30%, for several incorporation ratios between 5% and 30%. This reduction of workability is mainly related to the differences in fineness between cement and binder wastes (as forest biomass ashes, wood ash, cork ash, olive ash, paper-pulp ash, among others) and also differences on texture, shape, roughness, and water absorption. In general, waste binders present higher surface area and higher absorption, which decreases workability. Fig. 4.2 shows that the higher the volume of incorporation the higher the influence on workability.

4.2 Water/binder ratio

In render applications, the amount of water required depends on the workability of the mortars. This amount is the one that provides a proper workability. Part of this water does not participate in the chemical reactions of the hydraulic binders and evaporates after application, giving origin to voids in the interior of the mortars' matrix. These air voids are responsible for the porosity of the mortars, which influences their hardened properties such as deformability and mechanical strength.

The water/binder ratio is the first characteristic to be modified when the initial particles that compose a render are replaced by others, such as waste particles. This property of the mortars is sensitive to changes such as shape, roughness, and porosity. In Table 4.1, several examples of water/binder ratio of render mortars with different waste contents and for different waste types are summarized. In the table, only studies with workability similar between the reference mortar (0% wastes) and the mortars modified by wastes incorporation are presented.

In Fig. 4.3, the relative ratio between the modified mortars' water/binder ratio and the reference mortar (with no wastes) is represented for several mortars with different types of incorporation. In this figure, only studies that defined the water/binder ratio for a given workability are presented. There are several studies that incorporate wastes with the same water/binder ratio in mortars, but the modified mortars do not present proper workability. In those studies, it is necessary to add some admixtures, or they

Table 4.1 Water/binder ratio of mortars with incorporation of wastes where workability was maintained constant.

Type of incorporation	Waste	0%	5%	10%	15%	20%	40%	50%	100%	References
Aggregate	Red ceramic	0.96	-	-	-	1.00	-	1.17	1.50	[221]
	Sanitary ware	1.14	-	-	-	1.20	-	1.25	1.35	[18]
	Concrete	1.21	-	-	-	1.31	-	1.30	1.27	[8]
		0.94	0.96	0.98	-	1.20	1.07	-	-	[9]
	Construction and demolition waste (predominantly concrete)	1.15	-	-	-	1.21	-	1.33	1.44	[155]
	Construction and demolition waste (mixture)	1.15	-	-	-	1.13	-	1.19	1.40	[155]
	Glass	1.19	-	-	-	1.19	-	1.19	1.19	[143]
		0.60	-	-	-	-	-	0.56	0.47-0.50	[141]
	Plastics (PET (Polyethylene terephthalate) flakes)	1.36	1.36	1.36	1.36	-	-	-	-	[137]
	Plastics (PET pellets)	1.36	1.36	1.36	1.36	-	-	-	-	[137]
Filler	Red ceramics powder	0.96	0.86	0.86	-	-	-	-	-	[49]
	Sanitary ware powder	1.14	-	1.01	0.97	0.94	-	-	-	[7]
	Concrete powder	1.41	1.24	1.18	1.12	-	-	-	-	[51]
	Construction and demolition waste powder (predominantly concrete)	1.12	-	1.01	0.99	0.96	-	-	-	[154]
	Construction and demolition waste powder (mixture)	1.12	-	0.97	0.96	1.00	-	-	-	[154]
	Glass powder	1.41	-	1.24	1.21	1.19	-	-	-	[142]
	GFRP (Glass Fiber Reinforced Polymer) powder	1.30	-	0.90	0.90	0.90	-	1.00	-	[172]
Binder	Forest biomass ashes	1.30	1.30	1.30	1.40	-	-	-	-	[215]
		0.55	-	0.55	-	0.6	-	-	-	[182]
		0.55	-	0.55	-	0.55	-	-	-	[182]
	Fluid catalytic cracking	1.30	1.20	1.30	1.30	-	-	-	-	[218]
Fibre	Textile fibres (1.5 cm long)	1.00	-	-	-	-	-	-	-	[218]
	Textile fibres (3.0 cm long)	1.30	1.40	-	-	-	-	-	-	[218]
	Acrylic fibres (1.5 cm long)	1.30	1.40	-	-	-	-	-	-	[218]
	Acrylic fibres (3.0 cm long)	1.30	1.30	-	-	-	-	-	-	[218]
	Glass fibres (1.5 cm long)	1.30	1.40	-	-	-	-	-	-	[218]
	Glass fibres (3.0 cm long)	1.30	1.30	-	-	-	-	-	-	[218]
	Wool fibres (1.5 cm long)	1.30	1.40	-	-	-	-	-	-	[126]
	Wool fibres (3.0 cm long)	0.91	-	0.92	-	0.89	-	-	-	[126]
	Wool fibres (1.5 cm long)	0.91	-	0.98	-	0.89	-	-	-	[126]
	Wool fibres (3.0 cm long)	0.98	-	1.02	-	0.97	-	-	-	[126]

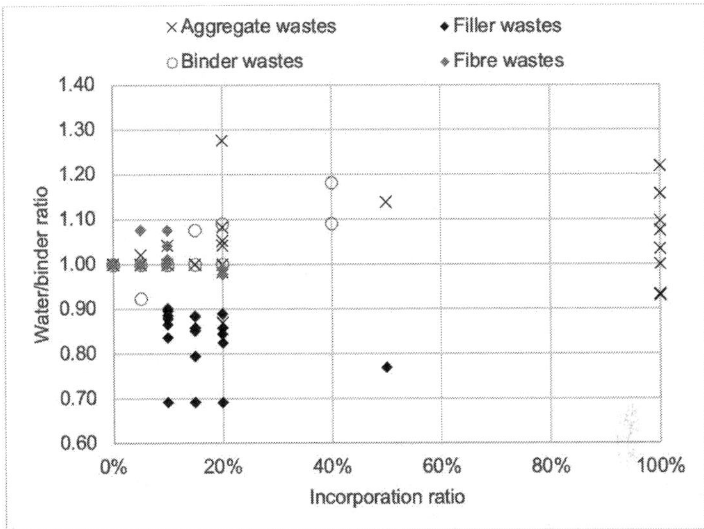

Fig. 4.3 Water/binder ratio of mortars with aggregate, filler, binder, or fiber wastes [7]–[9] [18] [49] [51] [126] [137] [141] [142] [154] [155] [172] [215] [218].

cannot be applied on vertical façades and are only theoretical approaches that cannot be applied in practice.

Similar to a workability analysis, in terms of water/binder ratio (Fig. 4.3), it is noticed that the type of incorporation has a preponderant influence. Filler wastes are responsible for a decrease in water needs. As mentioned above, these thin particles confer a good plasticity to the mortars, decreasing the amount of water required in the pastes. In some cases, the reduction can reach 30%. That was the case of the incorporation of a polymer filler GFRP [172] that conferred a great workability to the render, decreasing significantly the water needs of these type of mortars.

On the contrary, in general, the incorporation of aggregate wastes in mortars increases the amount of water needed to lubricate the surface of the wastes and to achieve a proper workability, thus the mortars present a higher water/binder ratio. An increase of 30% was noticed by Ledesma et al. [9] with concrete wastes aggregates.

Binder wastes also increase the water needs of the mortars. As mentioned above, this type of incorporation has a negative effect on workability, needing a higher volume of water to achieve a suitable plasticity. An increment of 18% in water/binder ratio was found by Rajamma et al. [182] when analyzing the replacement of cement with forest biomass ashes.

Fiber reinforcement in render mortars has a negative effect on workability and consequently increases the water needs of the mortars. The fibers absorption and adsorption are responsible for worsening workability and consequently increasing the water/binder ratio. However, the increase is in general lower than 10% (Fig. 4.3).

4.3 Bulk density

The bulk density is another characteristic analyzed in the fresh state tests. It is a relevant characteristic because, in the fresh state, it provides an indication about the possible porosity and compactness that the mortar will have in the hardened state. The bulk density is mainly influenced by the bulk density of wastes particles and their packing in the mortars. In Table 4.2, the fresh bulk density of mortars with aggregate, filler, binder, and fiber wastes from several studies is presented.

In Fig. 4.4, the relative bulk density, between modified mortars (by the incorporation of wastes) and the reference mortars, of several studies is presented. The relative bulk density varies between plus 5% and minus 15%, meaning that the incorporation of wastes can reduce this density up to 15% or increase it up to 5%, depending on the waste incorporated. In fact, the incorporation of wastes does not significantly affect the mortars' bulk density.

In general, filler wastes increase the bulk density of mortars made with them, and aggregate wastes decrease it, due to the rearrangement of the particles. In some cases, the bulk density of the waste particles is responsible for modifying these trends. For example, the incorporation of GFRP powder waste [172] would theoretically increase the bulk density due to a "filler effect", but as the polymer particles present a lower bulk density the fresh density of the paste is lowered.

The bulk density of binder wastes and fibers is noticed to be higher than the correspondent reference mortars in Fig. 4.4. This increase of bulk density is related to the volume of water of this fresh paste, when compared with the reference mortar. These wastes retain more water, increasing the mortars' bulk density in the fresh state.

4.4 Water retention

The water retention is the capacity of the mortars to retain their mixing water, when in contact with a porous substrate. The amount of water absorbed by the substrate can be higher or lower, according to the substrate's porosity and the surface's humidity. The water absorption by the substrate can hinder the cement hydration, and the mortar may not achieve total hydration. In Table 4.3, the water retention of several studies that analyzed this property in mortars with wastes incorporation is presented.

The mortars with wastes' incorporation, in general, present greater water retention as noticed in Fig. 4.5. The wastes' particles are responsible for hindering the flow of the mixing water to the substrate. This retention is due to the often higher specific surface area, higher porosity, and higher roughness of the wastes when compared to natural aggregates or to standard binders. In this sense, mortars with wastes' incorporation retain more water when in contact with a porous substrate that can be essential to cement hydration, improving this property of the mortars.

Table 4.2 Bulk density of mortars with incorporation of wastes (kg/m^3).

Type of incorporation	Waste	0%	5%	10%	15%	20%	50%	100%	References
Aggregate	Red ceramic	2066	-	-	-	1968	1882	1757	[221]
	Sanitary ware	2028	-	-	-	1981	1917	1780	[18]
	Concrete	1923	-	-	-	1910	1863	1720	[8]
	Construction and demolition waste (predominantly concrete)	1952	-	-	-	1910	1859	1770	[155]
	Construction and demolition waste (mixture)	1952	-	-	-	1838	1765	1698	[155]
	Glass	1920	-	-	-	1886	1813	1758	[143]
	Shredded tires	1942	1861	1802	1748	-	-	-	[148]
	Plastics (PET flakes)	1912	1878	1828	1795	-	-	-	[137]
	Plastics (PET pellets)	1912	1884	1852	1845	-	-	-	[137]
Filler	Red ceramics powder	2066	2021	1989	-	-	-	-	[49]
	Sanitary ware powder	2028	-	2034	2058	2064	-	-	[7]
	Concrete powder	1923	1930	1947	1975	-	-	-	[51]
	Construction and demolition waste powder (predominantly concrete)	1952	-	1973	1980	1982	-	-	[154]
	Construction and demolition waste powder (mixture)	1952	-	1862	1858	1908	-	-	[154]
	Glass powder	1930	-	1890	1880	1900	-	-	[142]
	GFRP powder	1975	-	1735	1749	1786	1901	-	[172]
Binder	Forest biomass ashes	1975	2058	2061	2045	-	-	-	[215]
		1728	-	-	-	-	-	1775	[180]
		1728	-	-	-	-	-	1755	[180]
		3170	-	3060	-	3100	3110	-	[182]
		3170	-	2940	-	3000	3170	-	[182]
	Fluid catalytic cracking	1975	2041	2022	2006	-	-	-	[218]
Fibre	Textile fibres (1.5 cm long)	1975	2027	-	-	-	-	-	[218]
	Textile fibres (3.0 cm long)	1975	2033	-	-	-	-	-	[218]
	Acrylic fibres (1.5 cm long)	1975	2028	-	-	-	-	-	[218]
	Acrylic fibres (3.0 cm long)	1975	2043	-	-	-	-	-	[218]
	Glass fibres (1.5 cm long)	1975	2044	-	-	-	-	-	[218]
	Glass fibres (3.0 cm long)	1975	2054	-	-	-	-	-	[218]
	Wool fibres (1.5 cm long)	2005	-	1978	-	1937	-	-	[126]
	Wool fibres (3.0 cm long)	2005	-	1970	-	2003	-	-	[126]
	Wool fibres (1.5 cm long)	1999	-	1980	-	1970	-	-	[126]
	Wool fibres (3.0 cm long)	1999	-	1978	-	1986	-	-	[126]

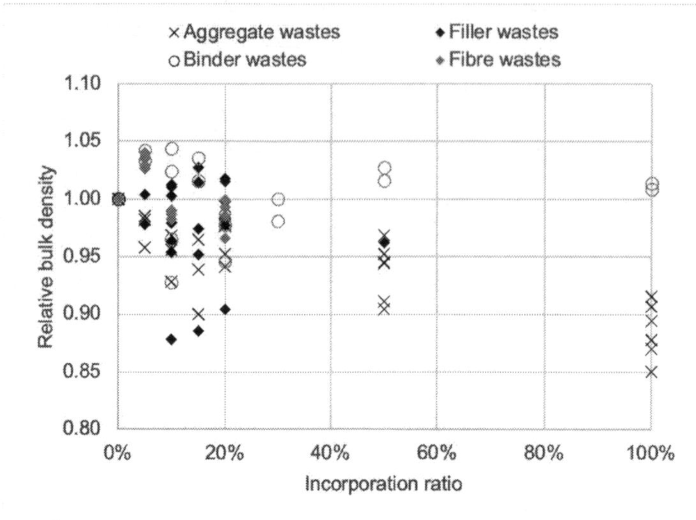

Fig. 4.4 Bulk density of mortars with aggregate, filler, binder, or fiber wastes [7] [8] [18] [49] [51] [126] [137] [142] [143] [148] [154] [155] [172] [180] [182] [215] [218].

Table 4.3 Water retention of mortars with incorporation of wastes (%).

Type of incorporation	Waste	0%	5%	10%	15%	20%	50%	100%	References
Aggregate	Red ceramic	72.2	-	-	-	-	78.4	-	[221]
	Sanitary ware	72.7	-	-	-	-	75.4	-	[18]
		66.6	-	-	-	-	-	74.2	[218]
	Concrete	63.8	-	-	-	89.3	-	-	[8]
	Glass	69.5	-	-	-	71.9	78.6	83.3	[143]
	Plastics (PET flakes)	74.9	-	-	73.7	-	-	-	[137]
	Plastics (PET pellets)	74.9	-	-	73.1	-	-	-	[137]
Filler	Red ceramics powder	72.2	-	85.1	-	-	-	-	[49]
	Sanitary ware powder	72.7	-	-	-	82.0	-	-	[7]
	Concrete powder	63.8	-	-	79.6	-	-	-	[51]
	Glass powder	63.2	-	-	-	78.9	-	-	[142]
	GFRP powder	66.6	-	-	-	-	87.2	-	[218]
Binder	Forest biomass ashes	66.6	-	-	74.1	-	-	-	[218]
		97.5	-	-	-	-	96.2	-	[180]
		97.5	-	-	-	-	95.0	-	[180]
Fibre	Textile fibres (1.5 cm long)	66.6	76.4	-	-	-	-	-	[218]

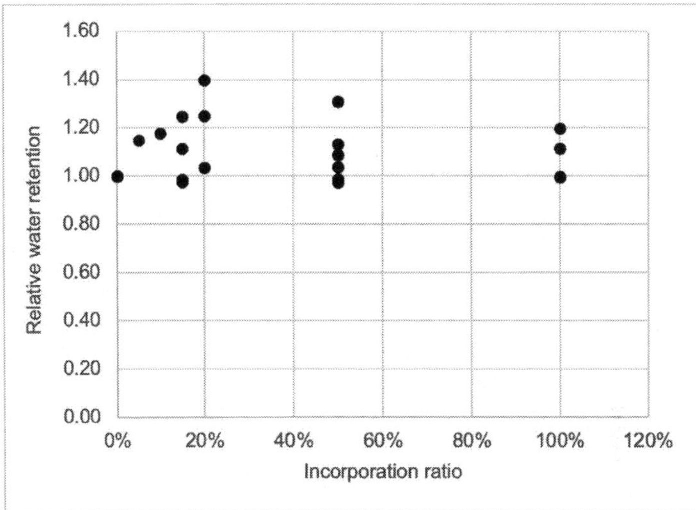

Fig. 4.5 Water retention of mortars with aggregate, filler, binder, or fiber wastes [7] [8] [18] [49] [51] [137] [142] [143] [180] [218].

4.5 Chapter conclusions

Adequate fresh properties in render mortars are essential. This specific type of mortars should easily be transported, applied, and made to adhere properly to the substrate's surface, covering its irregularities and easy spreading on all the façade's surface. In other types of mortars, such as masonry or concrete repair mortars, these characteristics may have not the same importance as in rendering mortars. Without a suitable plasticity, the mortars cannot be applied on vertical façades.

In this chapter, the influence of wastes on the mortars' fresh properties was analyzed. Their rough surface, irregular shape, and porosity are responsible for the changes in the mortars' fresh behavior. However, each waste influences differently these fresh properties. Filler wastes, incorporated in the mortars as fillers, improve the workability of the mortars, increase bulk density, and retain more water than a reference mortar. On the contrary, the wastes incorporated as aggregates, binders, or fibers, in general, are responsible for more consistent pastes, and consequently it is necessary to add a higher volume of water to the mixes to obtain a similar workability.

Mortars with incorporation of wastes: mechanical behavior

5

5.1 Mechanical strength

Thermal variations, humidity variations, ice formation, movements of the substrate, human blows, impact of objects are examples of actions that render mortars can be submitted to during their lifetime. The mortars should resist these environment or human actions and others. To resist these events, a render mortar should have an adequate mechanical strength. In this section, the influence of wastes' incorporation in mortars' mechanical behavior is approached, considering compressive strength, tensile strength, adherence strength, and impact resistance.

5.1.1 Compressive strength

Render mortars are not usually submitted to significant compressive stresses. Other mortar types as masonry or concrete reinforced repair mortars are more susceptible to this issue due to the vertical loads they are subjected to. Render mortars, on the other hand, are applied on vertical façades to which they adhere and are generally not subjected to vertical stresses generating compressive strength.

Compressive strength, however, gives an indication of the overall mechanical strength of the mortar; hence, its ability to withstand impacts, abrasion, and other aggressive mechanical actions that can compromise the durability of mortars.

Renders' compressive strength can be analyzed in laboratory by applying an increasing load (Fig. 5.1), registering their deformation, until failure is achieved.

When submitted to a compressive load, a mortar first presents an elastic phase; then for higher load incidence, the microstructure of the mortar begins to damage, starting a plastic phase [222]. The mortar's microstructure accommodates that stress, reducing the diameter of voids, until the particles start compressing and failure is achieved. Mortars' compressive strength is mainly dependent on their porosity volume and distribution. Porous materials present a lower compressive strength than more compact ones. The compression strength creates stresses in the matrix and over the voids, reducing them and decreasing the mortar's strength. Wastes' incorporation in a mortar modifies its microstructure, which is also responsible for a modification in compressive strength.

Aggregate wastes are more porous than conventional natural ones. The increment of porosity, when compared with a natural aggregate, is responsible for higher water absorption and consequently can improve the aggregate/matrix bond [82]. The waste can absorb a mixture of water and binder mechanically strengthening this bond. On the other hand, the higher contact area between the aggregate waste and the matrix can also strengthen the aggregate/matrix bond [82]. In Table 5.1, some examples of compressive

Eco-efficient Rendering Mortars. DOI: 10.1016/C2018-0-04147-X

Fig. 5.1 Compressive strength test illustration.

Table 5.1 Compressive strength of blended mortars with aggregate wastes' incorporation obtained by several authors.

Waste	Incorporation ratio (in volume)	Compressive strength (MPa)			References
		28 days	90 days	120 days	
Red ceramics	20%	8.97 (+13%)	-	-	[221]
	50%	9.11 (+15%)	-	-	
	100%	6.92 (−13%)	-	-	
Sanitary ware	20%	6.79 (+48%)	4.40 (+10%)	4.25 (+28%)	[18]
	50%	8.04 (+76%)	4.87 (+22%)	4.24 (−2%)	
	100%	5.85 (+28%)	3.93 (−2%)	3.47 (−20%)	
Concrete	20%	5.34 (+37%)	-	-	[8]
	50%	5.15 (+32%)	-	-	
	100%	7.38 (+89%)	-	-	
CDW - Construction and demolition wastes (predominantly concrete)	20%	6.44 (+54%)	3.73 (−10%)	-	[155]
	50%	5.98 (+43%)	3.31 (−20%)	-	
	100%	6.43 (+54%)	3.29 (−20%)	-	
CDW - Construction and demolition wastes	20%	4.22 (+1%)	4.03 (−2%)	-	[155]
	50%	3.00 (−29%)	2.40 (−58%)	-	
	100%	3.76 (−10%)	2.09 (−50%)	-	
Glass	20%	6.72 (+58%)	6.08 (83%)	-	[143]
	50%	7.20 (+70%)	7.04 (+111%)	-	
	100%	5.72 (+35%)	5.97 (+79%)	-	
Shredded tires	5%	3.84 (−9%)	3.81 (−23%)	-	[148]
	10%	2.73 (−35%)	3.10 (−38%)	-	
	15%	2.52 (−40%)	2.67 (46%)	-	
PET (Polyethylene Terephthalate) flakes	5%	3.58 (−3%)	3.31 (−13%)	-	[137]
	10%	1.91 (−49%)	1.94 (−49%)	-	
	15%	1.84 (−51%)	1.89 (−51%)	-	
PET (Polyethylene Terephthalate) pellets	5%	3.31 (−11%)	3.11 (−19%)	-	[137]
	10%	3.30 (−11%)	3.12 (−18%)	-	
	15%	2.96 (−20%)	2.91 (−85%)	-	

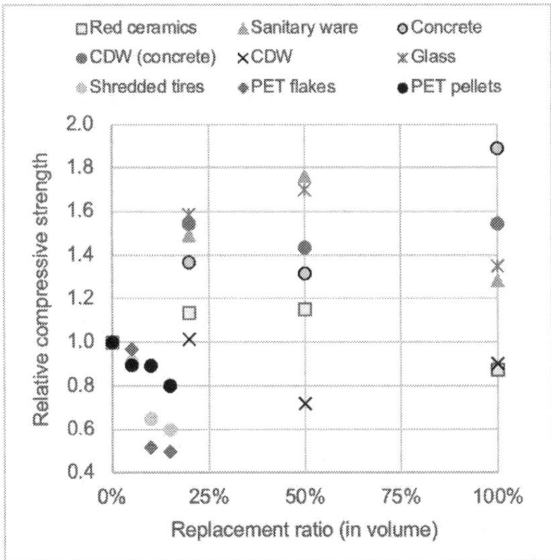

Fig. 5.2 Relative compressive strength between mortars with aggregate wastes and the corresponding plain mortar (28 days) [8] [18] [137] [143] [148] [155] [221].

strength of render mortars (1:4 cement: aggregate ratio) with aggregate wastes' incorporation are summarized. In Fig. 5.2, the relative compressive strength between these modified mortars and the correspondent plain mortar is also illustrated. From the examples, it is noticed that red ceramics (illustrated in Fig. 5.3), sanitary ware, glass, and concrete waste aggregates increase the mortars' compressive strength. Concrete waste, replacing the natural aggregate at 100%, increased this strength by more than 80% [8]. Sanitary ware also achieved strength of 80% higher than their corresponding plain mortar, when replacing the natural aggregate at 50% (in volume) [18].

Fig. 5.3 Illustration of red ceramics waste from bricks.

Not all wastes are porous or present a high specific surface area. Polymers, for example are not porous or absorbent materials. The characteristics of these types of wastes hinder proper aggregate/matrix bond. Shredded tires, PET flakes, or PET pellets are examples of wastes that are responsible for a decrease of compressive strength (Table 5.1 and Fig. 5.2). PET flakes, replacing the natural aggregate at 15% (in volume) reduced the mortars' compressive strength by 50% [137]. This reduction can be due to the bond matrix/aggregate or due to this type of aggregates' mechanical characteristics, as these aggregates are weaker than conventional ones.

Filler wastes, on the other hand, are in general responsible for increments of strengths. These powder wastes, replacing part of the natural aggregate, fill part of the empty voids between the particles and are responsible for more compact matrixes. In addition, these types of wastes, as discussed in Chapter 4, usually need less water to achieve a similar workability (to the reference mortar's). Consequently, in this type of mortars, the water/binder (w/b) ratio is lower than in the correspondent reference mortar. A lower w/b ratio is responsible for less porous matrix and consequently stronger mortars.

The increment of strength and the reduction of water absorption are the best advantages of using filler wastes. Table 5.2 summarizes the compressive strength of several filler wastes' incorporation in mortars. The relative compressive strength between these wastes and the mortar without wastes is illustrated in Fig. 5.4. Similar to aggregate wastes' analysis, the examples illustrated in the captions are from render mortars with a volumetric ratio of 1:4 (binder: aggregates) and the binder is CEM II/B-L 32.5 N. From these examples, it can be verified that all mortars with wastes' incorporation showed increments in strength. Sanitary ware powder and CDW (predominantly composed by concrete) powder were the filler wastes that most incremented the compressive strength. At 28 days, sanitary ware powder (20%) incremented the compressive strength threefold compared to the corresponding plain mortar, 14.92 MPa versus 4.56 MPa [7]. Mortars with filler wastes present compressive strengths considerably higher than those of conventional mortars. This allows reducing the amount of cement of mortars to decrease their strength and improve their deformability, originating mortars with lower incorporated energy.

From Tables 5.3–5.7, several examples of binder wastes' incorporation are summarized, namely biomass ashes: as forest biomass ashes (illustrated in Fig. 5.5), wood pellet ashes, eucalyptus bark ashes, cork ashes, olive tree ashes, and rice hush ashes. This type of wastes is designated "binder wastes" because they replace part of the initial binder as cement. The relative compressive strength between the blended mortars and the reference mortar is illustrated in Figs. 5.6 and 5.7. Some of the ashes presented are bottom and others are fly ashes. These two different types of ashes (bottom and fly) differ in several characteristics, namely in the particle size and specific surface area. Fly ashes are more reactive than bottom ashes, exactly due to their thinness when compared with bottom ashes. In the tables, several examples of treatments proposed by some authors are also presented, such as washing, gridding, combustion, or the elimination of light particles. In the tables, compressive strength test results by several authors are summarized. It is stressed that the studies presented in Tables 5.3– 5.7 with the exception of Farinha et al. [215] and Modolo et al. [180],

Table 5.2 Compressive strengths of blended mortars with filler wastes' incorporation obtained by other authors.

Waste	Incorporation ratio (in volume)	Compressive strength (MPa)						References
		28 days	90 days	120 days	180 days	365 days	730 days	
Red ceramics	5%	11.97 (+51%)	-	-	-	-	-	[49]
	10%	14.17 (+79%)	-	-	-	-	-	
Sanitary ware	10%	7.24 (+59%)	10.20 (+155%)	6.40 (+147%)	-	-	-	[7]
	15%	13.18 (+189%)	8.28 (+107%)	7.07 (+163%)	-	-	-	
	20%	14.92 (+227%)	9.55 (+139%)	7.77 (+179%)	-	-	-	
Concrete	5%	5.40 (+38%)	-	-	-	-	-	[51]
	10%	7.67 (+96%)	-	-	-	-	-	
	15%	8.64 (+121%)	-	-	-	-	-	
CDW (pre-dominantly concrete)	10%	5.99 (+44%)	6.12 (+49%)	-	-	-	-	[154]
	15%	12.06 (+189%)	7.19 (+75%)	-	-	-	-	
	20%	12.83 (+208%)	8.60 (+109%)	-	-	-	-	
CDW	10%	8.79 (+110%)	5.51 (+34%)	-	-	-	-	[154]
	15%	8.32 (+100%)	5.42 (+31%)	-	-	-	-	
	20%	7.77 (+86%)	4.53 (+110%)	-	-	-	-	
Glass	10%	5.55 (+31%)	5.02 (+51%)	-	-	-	-	[142]
	15%	6.81 (+61%)	6.96 (+109%)	-	-	-	-	
	20%	8.09 (+91%)	7.67 (+130%)	-	-	-	-	
GFRP	10%	5.92 (+14%)	6.11 (+71%)	-	5.99 (+51%)	6.71 (+15%)	6.83 (+14%)	[172]
	15%	6.38 (+23%)	6.93 (+94%)	-	7.35 (+86%)	8.16 (+39%)	8.06 (35%)	
	20%	7.15 (+38%)	9.33 (+161%)	-	8.86 (+124%)	10.96 (+87%)	10.90 (+82%)	
	50%	11.06 (+113%)	12.55 (+251%)	-	12.88 (+225%)	15.62 (+166%)	16.10 (+169%)	

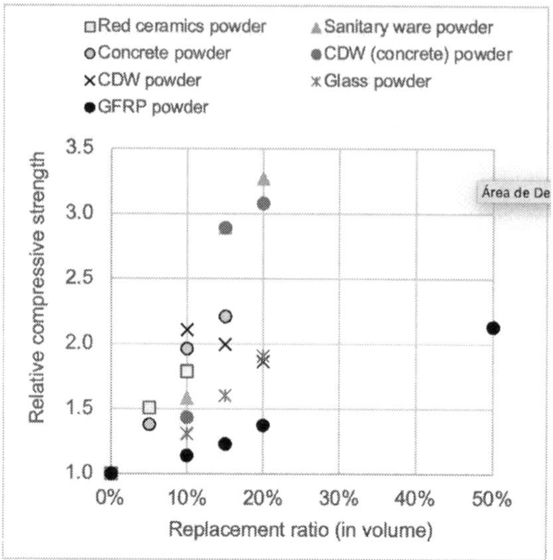

Fig. 5.4 Relative compressive strength between mortars with filler wastes and the corresponding plain mortar (28 days) [7] [49] [51] [142] [154] [172].

are not render mortars. The strengths obtained in the remaining studies are too high for render mortars, but are suitable for structural mortars, such as masonry mortars or reinforcement mortars. However, these are presented to indicate how binder wastes can affect mortars' compressive strength.

In general, the researchers noticed a decrease of strength with the replacement of cement with biomass ashes. For example, Rajamma et al. [55] analyzed compressive strengths at 28 and 365 days of mortars replacing CEM I 42.5R by 10%, 20%, and 30% (by weight) of forest biomass ashes. With the exception of the mortar with 10% of forest biomass at 28 days, all mortars at all ages presented lower compressive strength than the reference mortar. Great reduction of strength was found by several authors, such as Tosti et al. [53] when replacing CEM I 42.5R by 20% or 40% of wood pellets ashes or clean wood, cacao hush, and molasses ashes, or Ramos et al. [181] when replacing the binder with cork ash. Few studies found an increment of strength when biomass ashes replaced conventional Portland cement. A decrease of compressive strength due to the incorporation of binder wastes is expected, as they do not present the same hydraulic properties as standard binders. Some wastes can present some pozzolanicity, which can balance the decrease of strength produced by the cement replacement. However, the pozzolanic reaction occurs only at later ages. Additionally, the hydraulic compounds have in general different compositions, generating matrixes with lower strength. For example, Garcia and Sousa-Coutinho [179] analyzed the replacement of CEM I 42.5R with forest biomass ashes. In this research, the ashes incorporated were different from those used by Rajamma et al. [55] as they were bottom ashes. The authors replaced 10% cement with these ashes. A reduction

Table 5.3 Relative compressive strength between mortars with biomass ashes wastes and the corresponding plain mortar (Part 1 of 5).

Fuel	Type of ashes	Hydraulic element	Constituent replaced	Percentage of cement replacement	Compressive strength (MPa)	Age (days)	References
Forest wood and leaves	Bottom ashes	Rendering mortar	CEM II/B-L 32.5 N	5% (v.)	5.55 (+7%)	28	[215]
				10% (v.)	4.71 (−9%)		
				15% (v.)	4.30 (−17%)		
				5% (v.)	4.96 (+38%)	90	
				10% (v.)	4.43 (+23%)		
				15% (v.)	3.82 (+7%)		
				5% (v.)	4.41 (+11%)	180	
				10% (v.)	3.71 (−6%)		
				15% (v.)	2.92 (−26%)		
				5% (v.)	4.51 (−23%)	365	
				10% (v.)	3.69 (−37%)		
				15% (v.)	2.87 (−51%)		
				5% (v.)	4.24 (−30%)	730	
				10% (v.)	3.52 (−41%)		
				15% (v.)	3.32 (−44%)		
Forest wood	Fly ashes	Mortars	CEM I 42.5R	10% (wt.)	≈ 43 (+5%)	28	[55]
				20% (wt.)	≈ 33 (−20%)		
				30% (wt.)	≈ 22 (−46%)		
				10% (wt.)	≈ 43 (−10%)	90	
				20% (wt.)	≈ 41 (−15%)		
				30% (wt.)	≈ 38 (−20%)		
	Bottom ashes		CEM I 42.5R	5% (wt.)	47.5 (−4%)	28	[179]
				10% (wt.)	44.3 (−11%)		
				5% (wt.)	56.8 (−9%)	90	
				10% (wt.)	50.7 (−3%)		
				5% (wt.)	62.4 (+17%)	365	
				10% (wt.)	57.4 (+8%)		

Table 5.4 Relative compressive strength between mortars with biomass ashes wastes and the corresponding plain mortar (Part 2 of 5).

Fuel	Type of ashes	Hydraulic element	Constituent replaced	Percentage of cement replacement	Compressive strength (MPa)	Age (days)	References
Wood pellets	Fly ashes	Mortars	CEM I 42.5N	20% (wt.)	33.1 (−32%)	28	[53]
					≈ 37.8 (−33%)	90	
				40% (wt.)	21.0 (−57%)	28	
					≈ 26.5 (−53%)	90	
Clean wood, cacao hush, and molasses	Fly ashes	Mortars	CEM I 42.5N	20% (wt.)	46.7 (−4%)	28	
					≈ 47.3 (−17%)	90	
				40% (wt.)	30.0 (−37%)	28	
					≈ 33.1 (−42%)	90	
Eucalyptus bark	Fly ashes	Mortars	CEM I 42.5R	10% (wt.)	≈ 36 (−14%)	28	[55]
				20% (wt.)	≈ 30 (−29%)		
				30% (wt.)	≈ 28 (−33%)		
				10% (wt.)	≈ 38 (−20%)	90	
				20% (wt.)	≈ 34 (−30%)		
				30% (wt.)	≈ 34 (−30%)		
	No specified		CEM I 52.5 N/SR	10% (wt.)	54.5 (+19%)	28	[56]
				20% (wt.)	51.3 (+12%)		
				30% (wt.)	47.7 (+4%)		
Eucalyptus bark and pine bark	Fly ashes	Mortars	CEM I 42.5R	50% (wt.)	≈ 30 (−25%)	28	[57]
					≈ 36 (−12%)	56	
					≈ 38 (Equal)	90	
					≈ 46 (+4%)	180	
Cork	No specified	Mortars	CEM I 42.5R, No-grinded	10% (wt.)	51.2 (−10%)	28	[181]
				20% (wt.)	41.1 (−28%)		
			CEM I 42.5R, Grinded	10% (wt.)	45.2 (−17%)	28	
					56.0 (−5%)	90	
				20% (wt.)	39.4 (−27%)	28	
					46.8 (−21%)	90	

Table 5.5 Relative compressive strength between mortars with biomass ashes wastes and their plain mortar (Part 3 of 5).

Fuel	Type of ashes	Hydraulic element	Constituent replaced		Percentage of replacement	Compressive strength (MPa)	Age (days)	References
Oil palm	Fly ashes	Mortars	CEM I 52.5N		10% (wt.)	≈ 48 (−4%)	28	[60]
						≈ 55 (Equal)	90	
	No specified		CEM I	w/b = 0.50	10% (wt.)	48.4 (+5%)	28	[61]
						51.4 (+7%)	60	
						53.2 (+9%)	90	
					20% (wt.)	46.7 (+1%)	28	
						49.5 (+3%)	60	
						50.8 (+4%)	90	
					30% (wt.)	42.2 (−9%)	28	
						45.6 (−5%)	60	
						48.3 (−1%)	90	
					40% (wt.)	38.2 (−18%)	28	
						41.3 (−14%)	60	
						43.0 (−12%)	90	
				w/b = 0.575	10% (wt.)	44.3 (−1%)	28	
						46.9 (+0.2%)	60	
						49.2 (+2%)	90	
					20% (wt.)	44.0 (−2%)	28	
						47.4 (+1%)	60	
						49.5 (+2%)	90	
					30% (wt.)	40.1 (−11%)	28	
						43.6 (−7%)	60	
						46.2 (−5%)	90	
					40% (wt.)	35.8 (−20%)	28	
						38.8 (−17%)	60	
						40.2 (−17%)	90	
				w/b = 0.65	10% (wt.)	43.7 (+1%)	28	
						46.6 (+2%)	60	
						49.8 (+7%)	90	
					20% (wt.)	41.1 (−5%)	28	
						44.3 (−3%)	60	
						47.5 (+2%)	90	
					30% (wt.)	37.1 (−14%)	28	
						39.6 (−13%)	60	
						41.8 (−10%)	90	
					40% (wt.)	31.4 (−27.5)	28	
						34.3 (−25%)	60	
						36.4 (−22%)	90	
Paper sludge and wood biomass	Fly ashes	Mortars	CEM I 42.5N		20% (wt.)	45.6 (−6%)	28	[53]
						≈ 53.0 (−7%)	90	
					40% (wt.)	33.5 (−30%)	28	
						≈ 38.8 (−41%)	90	

Table 5.6 Relative compressive strength between mortars with biomass ashes wastes and the corresponding plain mortar (Part 4 of 5).

Fuel	Type of ashes	Hydraulic element	Constituent replaced	Treatment		Percentage of replacement	Compressive strength (MPa)	Age (days)	References
Olive tree	Bottom ashes	Mortars	CEM I 52.5	Noncrushed	Nontreated	20% (v.)	25.12 (−57%)	28	[63]
					Combusted		29.12 (−50%)		
					WLP		27.61 (−52%)		
				Crushed			36.18 (−38%)		
				Noncrushed	Nontreated		28.46 (−54%)	90	
					Combusted		35.21 (−43%)		
					WLP		32.05 (−48%)		
				Crushed			41.87 (−33%)		
				Noncrushed	Nontreated	38.5%	20.63 (−64%)	28	
					Combusted		23.88 (−59%)		
					WLP		27.66 (−52%)		
				Crushed			30.47 (−48%)		
				Noncrushed	Nontreated		32.61 (−43%)	90	
					Combusted		36.97 (−36%)		
					WLP		40.15 (−31%)		
				Crushed			47.28 (−19%)		
		Bedding mortars	CEM I 52.5N/SR			10% (v.)	52.06 (−11%)	28	[64]
						20% (v.)	45.54 (−22%)		
						10% (v.)	54.14 (−10%)	90	
						20% (v.)	47.69 (−21%)		
			CEM II/B-L 32.5N			10% (v.)	40.11 (−14%)	28	
						20% (v.)	29.74 (−36%)		
						10% (v.)	43.84 (−14%)	90	
						20% (v.)	32.98 (−35%)		

WLP, Without lightweight particles

Table 5.7 Relative compressive strength between mortars with biomass ashes wastes and the corresponding plain mortar (Part 5 of 5).

Fuel	Type of ashes	Hydraulic element	Constituent replaced		Percentage of replacement	Compressive strength (MPa)	Age (days)	References
Rice hush	Not specified	Mortars	CEM I	w/b = 0.50	10% (wt.)	47.3 (+2%)	28	[61]
						49.8 (+4%)	60	
						51.0 (+4%)	90	
					20% (wt.)	49.2 (+6%)	28	
						51.8 (+8%)	60	
						53.5 (+9%)	90	
					30% (wt.)	44.8 (−3%)	28	
						47.2 (−1.5%)	60	
						48.6 (−0.6%)	90	
					40% (wt.)	40.5 (−13%)	28	
						43.2 (+0.2%)	60	
						44.4 (−9.2%)	90	
				w/b = 0.575	10% (wt.)	45.7 (+2%)	28	
						50.4 (+8%)	60	
						53.6 (+11%)	90	
					20% (wt.)	49.2 (+10%)	28	
						52.9 (+13%)	60	
						55.6 (+15%)	90	
					30% (wt.)	43.6 (−3%)	28	
						46.1 (−2%)	60	
						48.0 (−0.8%)	90	
					40% (wt.)	38.1 (−15%)	28	
						42.1 (−10%)	60	
						43.8 (−10%)	90	
				w/b = 0.65	10% (wt.)	42.1 (−3%)	28	
						46.7 (+3%)	60	
						50.0 (+8%)	90	
					20% (wt.)	44.9 (+4%)	28	
						48.7 (+7%)	60	
						52.1 (+12%)	90	
					30% (wt.)	38.9 (−10%)	28	
						43.1 (−5%)	60	
						45.7 (−2%)	90	
					40% (wt.)	35.3 (−19%)	28	
						40.3 (−11%)	60	
						41.8 (−10%)	90	

in compressive strength until 90 days was also found by the authors. However, at 365 days, an increment of strength was noticed. A 5% replacement resulted in an increment of strength by about 18% and the replacement by 10% resulted in an increment of 8% in compressive strength. The late increment of strength can indicate that the wastes, studied by Garcia and Sousa-Coutinho [179], have a pozzolanicity potential responsible for an increment of strength.

Rosales et al. [63] analyzed the influence of olive tree biomass bottom ash with different treatments on cement-based mortars' compressive strength. The authors analyzed the biomass with and without treatment with crushed biomass,

Fig. 5.5 Illustration of a biomass forest bottom ash.

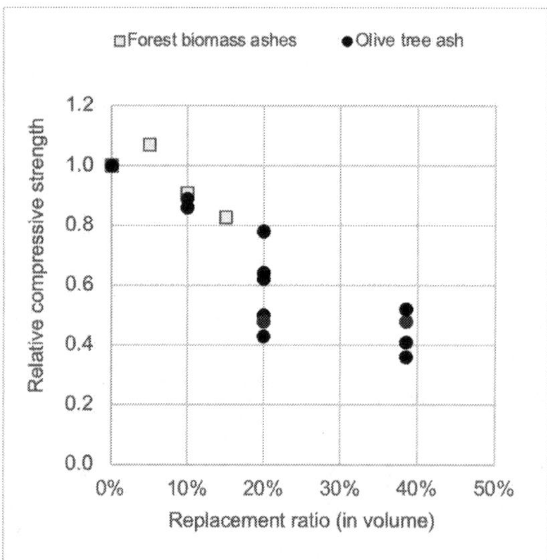

Fig. 5.6 Relative compressive strength between mortars with biomass ashes wastes and their plain mortar (28 days). [63] [64] [215].

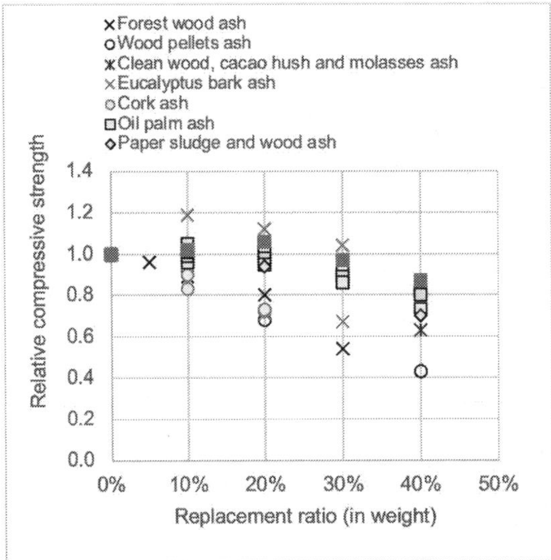

Fig. 5.7 Relative compressive strength between mortars with biomass ashes wastes and their plain mortar (28 days) [53] [55] [56] [57] [60] [61] [179] [181].

combusted biomass, and biomass without lightweight particles (WLP). Crushing was performed with a grinder and only particles finer than 125 μm were used. The combustion was a thermal process whereby the initial ashes were submitted to a temperature of 800 °C for 18 h. The elimination of lightweight particles was achieved by flotation. The different biomasses replaced the binder at 20% and 38.5% (by volume). The authors found that the mechanical behavior of the mortars was directly influenced by the incorporation ratio and the treatment process. The higher the biomass ratio the lower the mechanical strengths obtained. For compressive strength, 20% replacement resulted in a reduction between 33% and 57% and 38.5% replacement in a reduction between 29% and 57%. Ashes without treatment were the ones that presented worst mechanical behavior. On the contrary, crushed ashes were the ones that presented a better mechanical behavior (probably because fineness increments reactivity), within the treatments tested, followed by combusted ones and then by those WLP. The authors also analyzed the combination of the several treatments and found that this combination was beneficial.

In Tables 5.8 and 5.9, compressive strength test results for other type of binder wastes are presented, such as fluid catalytic cracking (FCC), a waste that is generated in the petroleum industry. Some authors evaluated the influence of this waste in the mortars' mechanical behavior. Relative compressive strength between the blended mortars and the corresponding plain mortar is illustrated in Fig. 5.8.

Table 5.8 Compressive strengths of blended mortars with FCC wastes' incorporation obtained by several authors (Part 1 of 2).

Hydraulic element	Constituent replaced	Percentage of replacement	Compressive strength (MPa)	Age (days)	References
Render mortars	CEM II/B-L 32.5 N	5%	4.25 (−18%)	28	[218]
		10%	3.64 (−30%)		
		15%	3.43 (−34%)		
		5%	4.58 (+28%)	90	
		10%	3.99 (+11%)		
		15%	3.97 (+11%)		
		5%	4.99 (+26%)	180	
		10%	3.92 (−1%)		
		15%	3.85 (−3%)		
		5%	3.20 (−45%)	365	
		10%	3.99 (+11%)		
		15%	3.97 (+11%)		
Mortars	CEM I 42.5R	5% (wt.)	33.4 (−6%)	2	[162]
			46.5 (Equal)	7	
			54.8 (+3%)	28	
			59.2 (+3%)	56	
			60.8 (+1%)	90	
		10% (wt.)	30.8 (−14%)	2	
			46.8 (+1%)	7	
			56.2 (+5%)	28	
			57.7 (+7%)	56	
			62.1 (+3%)	90	
		15% (wt.)	26.5 (−26%)	2	
			42.0 (−10%)	7	
			52.3 (−2%)	28	
			55.1 (−4%)	56	
			60.8 (+1%)	90	

		20% (wt.)			17.9 (−50%)	2	
					34.5 (−26%)	7	
					47.9 (−10%)	28	
					51.3 (−10%)	56	
					51.9 (−13%)	90	
	CEM I 52.5R	15% (wt.)	Noncarbonated	w/b = 0.3	105.7 (+12%)	70	[166]
				w/b = 0.4	91.3 (+25%)		
				w/b = 0.5	79.8 (+19%)		
				w/b = 0.7	52.9 (+1%)		
			Carbonated	w/b = 0.3	101.6 (+8%)		
				w/b = 0.4	94.3 (+21%)		
				w/b = 0.5	93.3 (+15%)		
				w/b = 0.7	69.6 (−7%)		
	CEM I	10% (wt.)	Without treatment		26.0 (−20%)	3	[169]
					34.3 (−5%)	7	
					40.4 (−7%)	28	
			Heated		28.0 (−13%)	3	
					40.4 (+12%)	7	
					47.1 (+7%)	28	
			Heated and grinded		31.9 (Equal)	3	
					43.1 (+20%)	7	
					49.5 (+13%)	28	

Table 5.9 Compressive strengths of blended mortars with FCC wastes' incorporation obtained by several authors (Part 2 of 2).

Hydraulic element	Constituent replaced	Percentage of replacement		Compressive strength (MPa)			Age (days)	References
				w/b = 0.42	w/b = 0.485	w/b = 0.55		
Mortars	CEM I	Equilibrium catalyst (Ecat)	5% (wt.)	≈ 26 (Equal)	≈ 31 (+15%)	≈ 22 (Equal)	7	[161]
				≈ 33 (+8%)	≈ 38 (+9%)	≈ 33 (Equal)	14	
				≈ 31 (+8%)	≈ 44 (+14%)	≈ 35 (+6%)	28	
				≈ 36 (Equal)	≈ 45 (+10%)	≈ 39 (+5%)	56	
			10% (wt.)	≈ 22 (−15%)	≈ 31 (+15%)	≈ 17 (−21%)	7	
				≈ 29 (−6%)	≈ 38 (+15%)	≈ 25 (−25%)	14	
				≈ 25 (−15%)	≈ 42 (+9%)	≈ 28 (−14%)	28	
				≈ 31 (−13%)	≈ 44 (+6%)	≈ 37 (+1%)	56	
			15% (wt.)	≈ 18 (−30%)	≈ 30 (+13%)	≈ 10 (−56%)	7	
				≈ 24 (−22%)	≈ 36 (+3%)	≈ 26 (−21%)	14	
				≈ 24 (−17%)	≈ 36 (−5%)	≈ 32 (−3%)	28	
				≈ 30 (−15%)	≈ 37 (−11%)	≈ 35 (−5%)	56	
		Electrostatic precipitator catalyst (EPcat)	5% (wt.)	≈ 29 (+16%)	≈ 33 (+22%)	≈ 30 (+1%)	7	
				≈ 35 (+10%)	≈ 36 (+22%)	≈ 37 (+8%)	14	
				≈ 41 (+15%)	≈ 40 (+17%)	≈ 35 (+6%)	28	
				≈ 42 (+10%)	≈ 42 (+14%)	≈ 40 (+1%)	56	
			10% (wt.)	≈ 29 (+20%)	≈ 35 (+27%)	≈ 25 (−15%)	7	
				≈ 36 (+13%)	≈ 37 (+24%)	≈ 32 (−7%)	14	
				≈ 39 (+9%)	≈ 41 (+22%)	≈ 28 (−16%)	28	
				≈ 41 (+6%)	≈ 44 (+18%)	≈ 35 (−11%)	56	
			15% (wt.)	≈ 28 (+12%)	≈ 33 (+22%)	≈ 21 (−28%)	7	
				≈ 33 (+3%)	≈ 39 (+32%)	≈ 27 (−22%)	14	
				≈ 34 (−5%)	≈ 45 (+33%)	≈ 27 (−20%)	28	
				≈ 34 (−11%)	≈ 48 (+30%)	≈ 34 (−14%)	56	

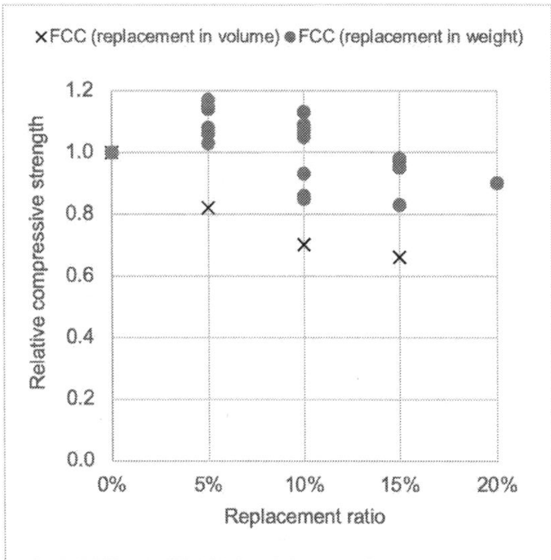

Fig. 5.8 Relative compressive strength between mortars with FCC wastes and the correspondent plain mortar (28 days) [162] [166] [169] [218].

Costa and Geada [162] found improvements when this waste is added to the mortars, until 10% of incorporation. Also Zornoza et al. [166] found increments of strength. Farinha et al. [215], who evaluated the compressive strength of render mortars, found a decrease of strengths at 365 days with the incorporation of this waste.

Render mortars, as already mentioned, do not have to present high compressive strengths; however, they should present an acceptable strength to resist the actions they are submitted to. The incorporation of binder wastes can reduce the mortars' compressive strength and this is not considered a drawback if an acceptable strength is guaranteed.

Textile, glass fibers from reinforced polymers (GFRP), nylon, or broom fibers are examples of fibers wastes that can reinforce the mortars. Generally, fibers absorb a higher amount of water than the remaining constituents of the mortars (natural aggregate or binder), as described in Chapter 3. This increment of water absorption can be beneficial to the fiber/matrix bond as the fibers can absorb a fresh mixture of water and binder. However, if the amount of water absorbed is too high, the opposite effect can occur. The water retained in the fibers' surface when evaporated creates voids/gaps between the fibers and the matrix, weakening the internal transition zone between the two materials. These empty spaces affect the compressive strength more than the flexural strength, as will be noticed in the next section.

The influence of these fibers wastes on mortars' compressive strength was analyzed by some authors and their results are summed up in Tables 5.10 and 5.11. The

Table 5.10 Compressive strengths of blended mortars with fibers wastes' addition obtained by several authors (Part 1 of 2).

Type of fiber	Addition percentage	Characteristics	Compressive strength (MPa) < 28 days	28 days	90 days	180 days	365 days	References
Sheep's wool	10%	L: 1.5 cm	-	8.91 (−8%)	-	-	-	[79]
		L: 3.0 cm	-	9.75 (+1%)	-	-	-	
	20%	L: 1.5 cm	-	8.62 (−11%)	-	-	-	
		L: 3.0 cm	-	12.19 (+26%)	-	-	-	
Textile	5%	L: 1.5 cm	2.50 (−25%)	4.53 (−13%)	4.52 (+28%)	4.33 (+9%)	3.32 (−43%)	[218]
		L: 3.0 cm	2.34 (−30%)	4.53 (−12%)	3.85 (+8%)	3.52 (−11%)	3.32 (−43%)	
	10%	L: 1.5 cm	-	8.97 (−7%)	8.81 (−8%)	-	-	[223]
		L: 3.0 cm	-	9.39 (−3%)	8.69 (−9%)	-	-	
	20%	L: 1.5 cm	-	6.17 (−36%)	7.28 (−24%)	-	-	
		L: 3.0 cm	-	8.02 (−17%)	7.86 (−18%)	-	-	
	1%	L: 2 cm	4.21 (+6%)	4.58 (+5%)	-	-	-	[176]
	4%	L: 2 cm	6.93 (+75%)	6.78 (+55%)	-	-	-	
	1%	L: 4 cm	3.76 (−5%)	3.80 (+13%)	-	-	-	
	4%	L: 4 cm	6.98 (+76%)	6.82 (+56%)	-	-	-	
	0.125%	-	-	4.14 (+10%)	-	-	-	[178]
	0.25%	-	-	3.77 (Equal)	-	-	-	
	0.50%	-	-	3.83 (+1%)	-	-	-	
GFRP	5%	L: 1.5 cm	2.60 (−22%)	5.27 (+2%)	4.93 (+37%)	4.64 (+17%)	4.63 (-21%)	[218]
		L: 3.0 cm	2.55 (−24%)	5.00 (−4%)	4.26 (+19%)	4.31 (+9%)	3.64 (−38%)	
	3%	Large	≈ 46 (+2%)	-	-	-	-	[174]
		Medium	≈ 31 (−31%)	-	-	-	-	
		Small	≈ 33 (−26%)	-	-	-	-	
		Powder	≈ 26 (−41%)	-	-	-	-	
	5%	EVE1	≈ 35 (−13%)	≈ 41 (−9%)	-	-	-	[190]
		EVE2	≈ 34 (−15%)	≈ 40 (−11%)	-	-	-	
		EVE3	≈ 34 (−15%)	≈ 40 (−11%)	-	-	-	
		UP (Unsaturated polyester)	≈ 31 (−23%)	≈ 39 (−13%)	-	-	-	
	0.25%	L: 2.0 cm	-	50.2 (+26%)	-	-	-	[202]
	0.75%		-	55.8 (+40%)	-	-	-	
	1.25%		-	59.2 (+48%)	-	-	-	

Table 5.11 Compressive strengths of blended mortars with fibers wastes' addition obtained by several authors (Part 2 of 2).

Type of fiber	Addition percentage	Characteristics		Compressive strength (MPa)					References
				< 28 days	28 days	90 days	180 days	365 days	
Acrylic	5%	L: 1.5 cm		2.83 (−15%)	5.08 (−2%)	4.64 (+30%)	3.84 (−3%)	4.08 (−30%)	[218]
		L: 3.0 cm		2.48 (−26%)	4.35 (−16%)	4.22 (+18%)	3.84 (−3%)	3.91 (−33%)	
Nylon	1%	L: 12.7 mm		-	32.7 (−34%)	-	-	-	[73]
		L: 25.4 mm		-	38.9 (−21%)	-	-	-	
		L: 38.1 mm		-	41.5 (−16%)	-	-	-	
	1.5%	L: 12.7 mm		-	31.3 (−37%)	-	-	-	
		L: 25.4 mm		-	36.3 (−27%)	-	-	-	
		L: 38.1 mm		-	39.6 (−20%)	-	-	-	
	1.0%	Straight	L: 20 mm	-	52.6 (−20.0%)	-	-	-	[74]
			L: 30 mm	-	53.6 (−18.4%)	-	-	-	
			L: 40 mm	-	55.5 (−15.4%)	-	-	-	
	1.5%		L: 20 mm	-	48.7 (−25.9%)	-	-	-	
			L: 30 mm	-	46.8 (−28.7%)	-	-	-	
			L: 40 mm	-	48.2 (−26.5%)	-	-	-	
	2.0%		L: 20 mm	-	34.1 (−48.0%)	-	-	-	
			L: 30 mm	-	35.1 (−46.6%)	-	-	-	
			L: 40 mm	-	35.3 (−46.3%)	-	-	-	
	0.5%	Knotted L: 40 mm		-	56.6 (−13.8%)	-	-	-	
	0.75%			-	56.0 (−14.9%)	-	-	-	
	1.0%			-	36.8 (−44.0%)	-	-	-	
Broom fibers	0.4%	L: 20 mm		-	≈ 42.5 (−5%)	-	-	-	[144]
		L: 50 mm		-	≈ 41.2 (−8%)	-	-	-	
	1.8%	L: 20 mm		-	≈ 40.3 (−10%)	-	-	-	
		L: 50 mm		-	≈ 43.0 (−4%)	-	-	-	

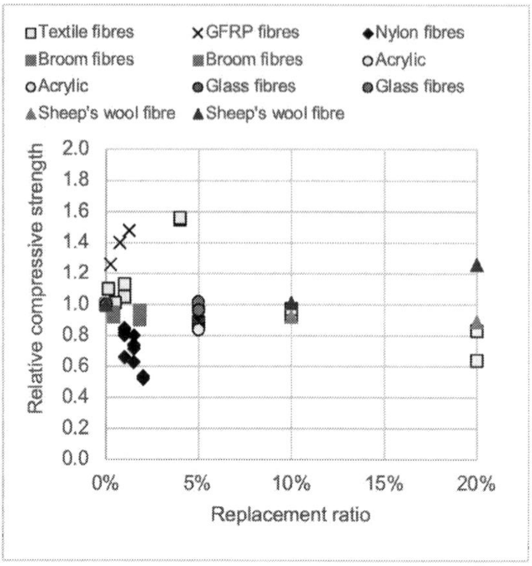

Fig. 5.9 Relative compressive strength between mortars with fiber wastes and their correspondent plain mortar (28 days) [74] [73] [126] [174] [176] [178] [190] [202] [218] [223].

relative values of compressive strength between these studies' results and the corresponding reference mortar are illustrated in Fig. 5.9.

From the captions, a general reduction of strength can be noticed. For example, Rodin et al. [174] evaluated four types of fibers (large, medium, small, and powder) recycled from wind turbine blades and then incorporated in cement mortars. In the compressive strength test (analyzed at 7 days), the authors found that only the large fibers presented a similar compressive strength to that of the reference mortar, as the remaining fibers were responsible for a reduction of strength between 26% and 41%. The incorporation of nylon fibers was investigated by Spadea et al. [73] and Orasutthikul et al. [74]. In the research, a reduction of compressive strength was noticed for all the incorporation contents of fibers as well as for all the fibers' lengths tested (12.7 mm, 20 mm, 25.4 mm, 30 mm, 38.1 mm, 40 mm). The broom fibers also reduced, although less significantly, the mortars' compressive strength, between 4% and 10% [144].

From the examples, it can be noticed that, of all the fibers, the ones that provide the mortars with better compressive strength are textile fibers and sheep's wool fibers. In addition, these fibers are the most absorbent ones, as they are a thread of thinner fibers, retaining water between the thinner fibers that compose the threads. This higher water absorption probably was favorable to the fiber/matrix bond, increasing the mortars' compactness, which resulted in higher compressive strength.

5.1.2 Tensile strength

Contrary to compressive stresses, to which render mortars are not usually subjected, they are constantly submitted to tensile ones. These are promoted by external actions

Fig. 5.10 Flexural strength test illustration.

as thermal stresses, variations of moisture, or movements of the substrate. They can also be promoted by internal actions as possible formation of ice and salts crystallization in the pores. Render mortars must accommodate these tensile stresses to avoid cracking, as will be discussed forward in this document.

The tensile strength of renders depends on mortar's matrix hardness and its ability to accommodate tensions. Usually tensile strength is evaluated by a flexural strength test (as illustrated in Fig. 5.10). The test consists in applying a load on a specimen supported in two points. The mortar will be tensioned in the lower face of the prism. The load is increased until the specimen reaches failure. The behavior of the specimen can be evaluated during the test by the tensile/deformation curves, as presented in Section 5.2.2.

Tensile strength is influenced by the mortars' matrix compactness, similar to compressive strength. Matrix's characteristics, aggregates/matrix bond, and mortars' porosimetry are all characteristics that influence mortars' tensile strength. Wastes influence these properties and, in this sense, they influence the mortars' tensile strength.

In Table 5.12, some examples of aggregate wastes' influence on flexural strength are summarized. The relative strength between these blended mortars and their reference mortar is illustrated in Fig. 5.11. The examples presented are from studies that replaced the natural aggregate with aggregate wastes by volume. The mortars have a volumetric ratio of 1:4 (binder: aggregate) and the wastes were incorporated with the same grain size distribution curve as the natural aggregate.

Similar to compressive strength, several wastes are responsible for increments of flexural strength when incorporated in mortars in replacement of natural aggregates. Red ceramics, sanitary ware, concrete, or glass are examples of wastes that can improve flexural strength. These increments of strengths are related to the physical properties of the wastes. These particles are rougher, present a higher specific surface area, and in some cases absorb a higher amount of water than natural aggregates. All these characteristics promote a stronger aggregate/matrix bond.

Table 5.12 Flexural strengths of blended mortars with aggregate wastes' incorporation obtained by several authors.

Type of fiber	Incorporation ratio	Flexural strength (MPa)		References
		28 days	90 days	
Red ceramics	20%	2.34 (+10%)	-	[221]
	50%	2.28 (+8%)	-	
	100%	2.01 (+10%)	-	
Sanitary ware	20%	1.65 (+33%)	1.98 (+25%)	[18]
	50%	1.76 (+42%)	1.96 (+24%)	
	100%	1.41 (+14%)	1.83 (+16%)	
Concrete	20%	1.86 (+37%)	-	[8]
	50%	1.63 (+20%)	-	
	100%	2.15 (+58%)	-	
CDW (predominantly concrete)	20%	1.15 (−23%)	1.11 (+3%)	[155]
	50%	1.31 (−13%)	0.91 (−15%)	
	100%	1.57 (+5%)	0.91 (−15%)	
CDW	20%	1.55 (+4%)	1.19 (+11%)	[155]
	50%	0.97 (−35%)	0.73 (−32%)	
	100%	0.79 (−47%)	0.60 (−44%)	
Glass	20%	2.70 (+61%)	2.28 (+58%)	[143]
	50%	2.71 (+62%)	2.70 (+88%)	
	100%	2.32 (+39%)	2.40 (+67%)	
Shredded tires	5%	1.11 (−22%)	1.11 (−34%)	[148]
	10%	1.19 (−6%)	1.30 (−23%)	
	15%	0.99 (−22%)	1.04 (−39%)	
PET flakes	5%	1.29 (−6%)	1.31 (−7%)	[137]
	10%	0.85 (−38%)	0.73 (−48%)	
	15%	0.73 (−47%)	0.76 (−56%)	
PET pellets	5%	1.29 (−6%)	1.26 (−10%)	[137]
	10%	1.22 (−11%)	1.24 (−12%)	
	15%	1.21 (−12%)	1.19 (−15%)	

For example, Penacho et al. [143] replaced the natural aggregate with glass waste and achieved mortars with increments of flexural strength (at 90 days) of 58–88%, depending on the volume of glass incorporated. The incorporation of glass by 50% was the dosage that presented the highest flexural strength. The replacement of natural aggregate with concrete waste, on the other hand, was more beneficial for 100% of incorporation, incrementing the strength by 58%, when compared with the reference mortar [8].

The incorporation of polymer wastes in mortars, replacing natural aggregate, on the other hand, resulted in weaker mortars. Shredded tires, PET flakes, and PET pellets are examples of these wastes (Fig. 5.11). For PET wastes, strength was lowered with the volume of PET incorporated [137]. The mortar with PET flakes achieved a reduction of 47% of flexural strength when the natural aggregate was replaced at

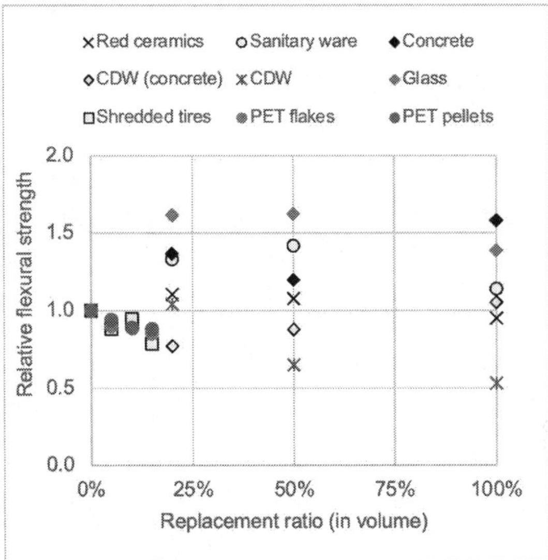

Fig. 5.11 Relative flexural strength between mortars with aggregate wastes and the correspondent plain mortar (28 days) [8] [137] [143] [148] [155] [221].

15%. These reductions of strength can be due to a worse fiber/matrix bond, probably due to the low water absorption of these wastes that do not favor this bond.

Filler wastes, as mentioned for compressive strength, are characterized by increments of strength. These fine particles, by filler effect, fill the voids between the natural aggregate grains and increment the mortars' compactness and consequently their mechanical strengths. Flexural strength is also incremented. In Table 5.13, examples of filler wastes' influence on flexural strength are presented. The relative increment of strength between these mortars with fillers and the correspondent reference mortar is illustrated in Fig. 5.12.

From Fig. 5.12, it is noticed that all examples of the mortars with filler wastes present higher flexural strength than the reference mortar (presenting a relative value higher than 1). Sanitary ware powder waste is the one, from these examples, that increments flexural strength the most. The waste, when incorporated as a filler at 20% (replacing the natural aggregate), increments the flexural strength by 250% [7], at 28 days. Concrete powder waste also significantly increments the mortars' flexural strength. Jesus et al. [154] found that CDW, composed of concrete powder, incremented the flexural strength by 150% when compared with their reference mortar.

Contrary to filler wastes, binder wastes present the opposite trend. This type of wastes, when replacing cement, reduces the mortars' mechanical strengths. Some examples of blended mortars with incorporation of biomass ashes as binder wastes are presented in Tables 5.14 and 5.15. The relative strengths between the blended mortars and their correspondent reference mortar are illustrated in Figs. 5.13 and 5.14. Similar to the mortars presented in the compressive strength analysis, the majority of

Table 5.13 Flexural strengths of blended mortars with filler wastes' incorporation obtained by several authors.

Type of fiber	Incorporation ratio	Flexural strength (MPa)					References
		28 days	90 days	180 days	365 days	730 days	
Red ceramics	5%	2.73 (+29%)	-	-	-	-	[49]
	10%	3.39 (+60%)	-	-	-	-	
Sanitary ware	10%	1.87 (+51%)	1.65 (4%)	-	-	-	[7]
	15%	2.74 (121%)	2.06 (+30%)	-	-	-	
	20%	4.39 (+254%)	2.27 (+44%)	-	-	-	
Concrete	5%	1.67 (+22%)	-	-	-	-	[51]
	10%	2.12 (+56%)	-	-	-	-	
	15%	2.71 (+99%)	-	-	-	-	
CDW (predominantly concrete)	10%	2.03 (+36%)	1.56 (+45%)	-	-	-	[154]
	15%	2.67 (+79%)	1.95 (+82%)	-	-	-	
	20%	3.54 (+138%)	2.01 (+88%)	-	-	-	
CDW	10%	2.96 (+98%)	1.52 (+42%)	-	-	-	[154]
	15%	3.04 (+104%)	1.54 (+44%)	-	-	-	
	20%	2.60 (+74%)	1.18 (+10%)	-	-	-	
Glass	10%	2.16 (+29%)	2.13 (+48%)	-	-	-	[142]
	15%	2.78 (+66%)	2.26 (+57%)	-	-	-	
	20%	3.10 (+86%)	2.71 (+88%)	-	-	-	
GFRP	10%	2.59 (+23%)	2.46 (+60%)	2.73 (+53%)	2.70 (+39%)	2.78 (+39%)	[172]
	15%	2.56 (+22%)	2.57 (+67%)	2.74 (+53%)	3.02 (+56%)	2.90 (+45%)	
	20%	2.65 (+26%)	2.57 (+67)	3.34 (+87%)	4.05 (+109%)	3.72 (+86%)	
	50%	3.13 (+49%)	3.05 (+98%)	3.46 (+93%)	4.95 (+155%)	4.57 (+129%)	

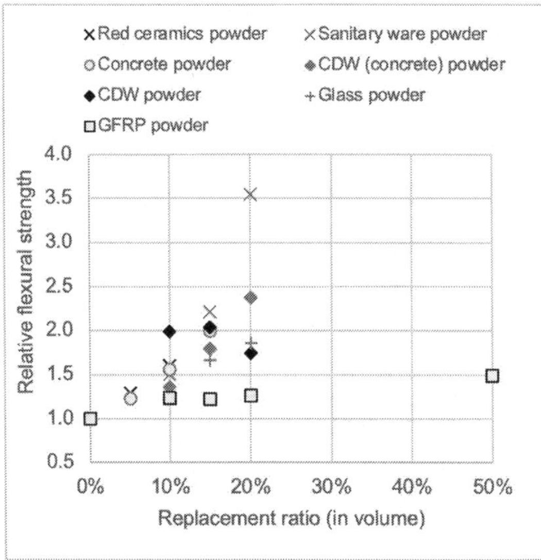

Fig. 5.12 Relative flexural strength between mortars with filler wastes and the correspondent plain mortar (28 days) [7] [49] [51] [142] [154] [172].

the mortars presented in the tables are not render mortars, with the exception of those from Farinha et al. [215].

In general, biomass ashes wastes reduce flexural strength. The reduction depends on the type of ash, its nature, and its physical and chemical characteristics. For example, Rosales et al. [63] investigated the influence of olive tree ashes' incorporation on mortars' flexural strength when replacing CEM I 52.5 at 20% and 38.1%. The authors found a decrease in strength for incorporation of 25%, resulting in reductions between 22% and 30%. Beltrán et al. [64] also analyzed the flexural strength of bedding mortars with olive tree ashes' incorporation replacing two different cement types (CEM I 52.5N/SR and CEM II/B-L 32.5N). For both cements, the authors found a decrease in strength between 14% and 35%.

There are, however, studies in which the biomass ashes were able to replace the standard binder and produce a similar strength. The research of Teixeira et al. [57] is an example of this. The authors incorporated forest biomass ashes and achieved similar mortars in terms of flexural strength.

Some authors investigated the influence that some treatments could have on the performance of the ashes in terms of the mortars' flexural strength. Ramos et al. [181] investigated the incorporation of cork ashes grinded and not-grinded and found that for longer ages the grinded ashes resulted in mortars with higher strengths those with not-grinded particles. Rosales et al. [63] also investigated the influence of several treatments in mortars' flexural strength: combusted, WLP and crushed. The ashes

Table 5.14 Flexural strengths of blended mortars with binder wastes' incorporation obtained by several authors (Part 1 of 2).

Fuel	Type of ashes	Hydraulic element	Constituent replaced	Percentage of replacement	Flexural strength (MPa)	Age (days)	References
Forest wood and leaves	Bottom ashes	Rendering mortar	CEM II/B-L 32.5 N	5% (v.)	1.89 (−10%)	28	[215]
				10% (v.)	1.80 (−14%)		
				15% (v.)	1.37 (−35%)		
				5% (v.)	1.79 (+16%)	90	
				10% (v.)	1.43 (−7%)		
				15% (v.)	1.49 (−3%)		
				5% (v.)	2.03 (+13%)	180	
				10% (v.)	1.46 (−18%)		
				15% (v.)	1.35 (−25%)		
				5% (v.)	2.14 (+10%)	365	
				10% (v.)	1.53 (−21%)		
				15% (v.)	1.38 (−29%)		
				5% (v.)	1.60 (−20%)	730	
				10% (v.)	1.46 (−27%)		
				15% (v.)	1.42 (−29%)		
Forest wood	Fly ashes	Mortars	CEM I 42.5R	10% (wt.)	≈ 3.0 (−3%)	28	[55]
				20% (wt.)	≈ 3.0 (−3%)		
				30% (wt.)	≈ 3.0 (−3%)		
				10% (wt.)	≈ 4.5 (−21%)	90	
				20% (wt.)	≈ 4.1 (−28%)		
				30% (wt.)	≈ 4.1 (−28%)		

	Bottom ashes		CEM I 42.5R	5% (wt.)	10.4 (-7%)	28	[179]
				10% (wt.)	10.1 (-10%)		
				5% (wt.)	11.1 (Equal)	90	
				10% (wt.)	10.4 (-7%)		
				5% (wt.)	9.4 (+18%)	365	
				10% (wt.)	8.9 (+11%)		
Eucalyptus bark	Fly ashes	Mortars	CEM I 42.5R	10% (wt.)	$\approx$ 3.0 (-5%)	28	[55]
				20% (wt.)	$\approx$ 3.0 (-5%)		
				30% (wt.)	$\approx$ 3.2 (Equal)		
				10% (wt.)	$\approx$ 4.2 (-26%)	90	
				20% (wt.)	$\approx$ 3.4 (-40%)		
				30% (wt.)	$\approx$ 3.0 (-47%)		
	No specified		CEM I 52.5 N/SR	10% (wt.)	8.4 (-2%)	28	[56]
				20% (wt.)	7.7 (-10%)		
				30% (wt.)	7.5 (-13%)		
Eucalyptus bark and pine bark	Fly ashes	Mortars	CEM I 42.5R	50% (wt.)	$\approx$ 6.4 (Equal)	28	[57]
					$\approx$ 7.9 (+5%)	56	
					$\approx$ 7.9 (+2%)	90	
					$\approx$ 8.1 (-4%)	180	
Cork	No specified	Mortars	CEM I 42.5R No-grinded	10% (wt.)	7.2 (-13%)	28	[181]
				20% (wt.)	6.4 (-23%)		
			CEM I 42.5R Grinded	10% (wt.)	6.6 (-13%)	28	
Palm oil	Fly ashes	Mortars	CEM I 52.5N	20% (wt.)	7.9 (-2%)	90	[60]
					6.0 (-21%)	28	
					6.9 (-15%)	90	
				10% (wt.)	$\approx$ 4.7 (-24%)	28	
					$\approx$ 5.4 (-19%)	90	

Table 5.15 Flexural strengths of blended mortars with binder wastes' incorporation obtained by other authors (Part 2 of 2).

Fuel	Type of ashes	Hydraulic element	Constituent replaced	Treatment		Percentage of replacement	Flexural strength (MPa)	Age	References
Olive tree	Bottom ashes	Mortars	CEM I 52.5	Noncrushed	Nontreated	20% (v.)	7.28 (−30%)	28 days	[63]
					Combusted		8.05 (−22%)		
					WLP		7.71 (−25%)		
				Crushed			8.01 (−22%)		
				Noncrushed	Nontreated		9.03 (−24%)	90 days	
					Combusted		9.32 (−22%)		
					WLP		9.28 (−22%)		
				Crushed			9.73 (−19%)		
				Noncrushed	Nontreated	38.5% (v.)	6.45 (−38%)	28 days	
					Combusted		6.74 (−35%)		
					WLP		6.59 (−36%)		
				Crushed			7.14 (−31%)		
				Noncrushed	Nontreated		7.48 (−37%)	90 days	
					Combusted		8.35 (−30%)		
					WLP		7.68 (−36%)		
				Crushed			8.62 (−28%)		
		Bedding mortars	CEM I 52.5N/SR			10% (v.)	9.69 (−17%)	28 days	[64]
						20% (v.)	9.31 (−21%)		
						10% (v.)	10.50 (−14%)	90 days	
						20% (v.)	9.79 (−20%)		
			CEM II/B-L 32.5N			10% (v.)	8.25 (−15%)	28 days	
						20% (v.)	6.25 (−35%)		
						10% (v.)	8.41 (−17%)	90 days	
						20% (v.)	6.74 (−34%)		

WLP, Without lightweight particles

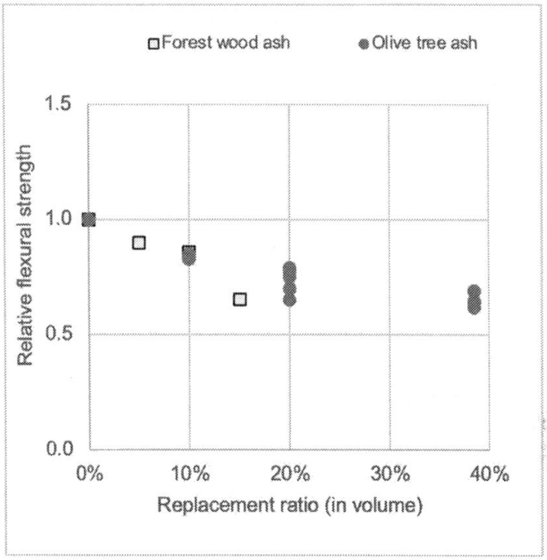

Fig. 5.13 Relative flexural strength between mortars with biomass ashes wastes and the corresponding plain mortar (28 days)—volume replacement [63] [64] [215].

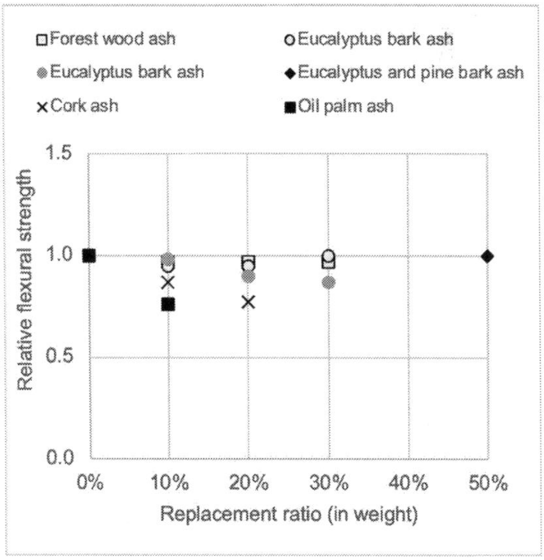

Fig. 5.14 Relative flexural strength between mortars with biomass ashes wastes and the corresponding plain mortar (28 days)—weight replacement [55] [56] [57] [60] [179] [181].

that were combusted or crushed were the ones that presented the highest flexural strength. The specific surface area and the particles grain size are characteristics that influence the chemical reactions. Finer particles are more reactive, and particles of higher specific surface area have also greater probability of reacting due to their contact area. Treatments can, in fact, improve some properties of the wastes, as discussed in Section 3.3. However, the treatments are responsible for several environmental impacts and must be carefully considered.

In Table 5.16, another type of binder wastes and their influence on flexural strength is presented. The waste FCC was in general responsible for a decrease in mortars' flexural strength (Fig. 5.15).

Farinha [218] investigated the replacement of FCC waste in render mortars, replacing Portland cement at 5%, 10%, and 15%, in volume. At 365 days, the waste was responsible for a reduction of strength of almost 50% for a replacement of 5% of the binder. This high reduction of strength is also due to the lower cement content of the original mortar. The initial mortar had a volumetric ratio of 1:4 (binder: aggregate), being poor in cement. The replacement of 5% with waste, in volume, resulted in major decreases of strength. Probably, if the mortar were richer in cement (presenting a higher volume of cement), the decrease would not have been so high. However, the reduction of strength is clear.

In fact, although the binder wastes can present some pozzolanicity, they do not have the same properties of a hydraulic binder as cement, being unable to achieve the same level of strength.

Fibers addition in mortars, on the other hand, can be responsible for an improvement in flexural strength. That is because fibers function as bridges in the cement matrix, delaying failure. This bridging effect is responsible for minimizing the effect of microcracks that are formed when a specimen is submitted to a tensile stress. With this bridging, the mortar can withstand a greater load until failure is achieved. The effectiveness of these "bridges" depends on the fibers and on the mortar's matrix. The fibers chosen should present a tensile strength higher than the mortars' matrix, to be able to withstand a load that the matrix is unable to. In addition, the fibers should present a modulus of elasticity higher than the matrix's, to mobilize them and distribute the tensions in the cement matrix and decrease cracking. If the fibers present a modulus of elasticity too low, they will deform without creating a barrier to cracking. In this sense, the effectiveness of the fibers depends on the matrix in which they are encapsulated, for which compatibility between the two materials is necessary.

In Tables 5.17 and 5.18, several examples of mortars reinforced by fiber wastes and their influence on flexural strength are summarized. Improvements in strength were found in some studies. Pinto et al. [176], for example, found that the addition of textile fibers until 4% (of binder volume) in rendering mortars was responsible for an increment of mortars' flexural strength. The authors analyzed two lengths (2 cm and 4 cm) and for both noticed an increment in strength. The higher the fibers' addition, the higher the value of strength obtained. Another example is Spadea et al.'s [73] research, in which the authors noticed that the flexural strength of all the mortars reinforced with nylon fibers (an illustration of this type of fiber waste is in Fig. 5.16) presented higher strengths in comparison with the reference mortar. The

Table 5.16 Flexural strengths of blended mortars with FCC wastes' incorporation obtained by several authors.

Hydraulic element	Constituent replaced	Percentage of replacement			Flexural strength (MPa)	Age (days)	References
Render mortars	CEM II/B-L 32.5 N	5%			1.42 (−32%)	28	[218]
		10%			1.26 (−40%)		
		15%			1.17 (−44%)		
		5%			1.58 (+3%)	90	
		10%			1.37 (−11%)		
		15%			1.16 (−25%)		
		5%			2.04 (+13%)	180	
		10%			1.60 (−11%)		
		15%			1.48 (−17%)		
		5%			1.05 (−46%)	365	
		10%			0.90 (−54%)		
		15%			0.91 (−53%)		
Mortars	CEM I 42.5R	5% (wt.)			8.2 (−10%)	2	[162]
					9.6 (−11%)	7	
					11.2 (−1%)	28	
					11.6 (−3%)	56	
					11.4 (−6%)	90	
		10% (wt.)			7.2 (−21%)	2	
					9.0 (−17%)	7	
					9.8 (−13%)	28	
					11.3 (−6%)	56	
					11.1 (−8%)	90	
		15% (wt.)			6.5 (−29%)	2	
					8.2 (−24%)	7	
					9.1 (−20%)	28	
					9.1 (−24%)	56	
					9.8 (−19%)	90	
		20% (wt.)			4.2 (−54%)	2	
					7.0 (−35%)	7	
					8.1 (−28%)	28	
					9.2 (−23%)	56	
					8.0 (−34%)	90	
	CEM I 52.5R	15% (wt.)	Noncarbonated	w/b=0.3	11.1 (−11%)	70	[166]
				w/b=0.4	11.0 (+3%)		
				w/b=0.5	11.1 (+1%)		
				w/b=0.7	8.0 (−21%)		
			Carbonated	w/b=0.3	9.7 (+9%)		
				w/b=0.4	10.9 (+17%)		
				w/b=0.5	7.2 (+20%)		
				w/b=0.7	8.7 (−16%)		

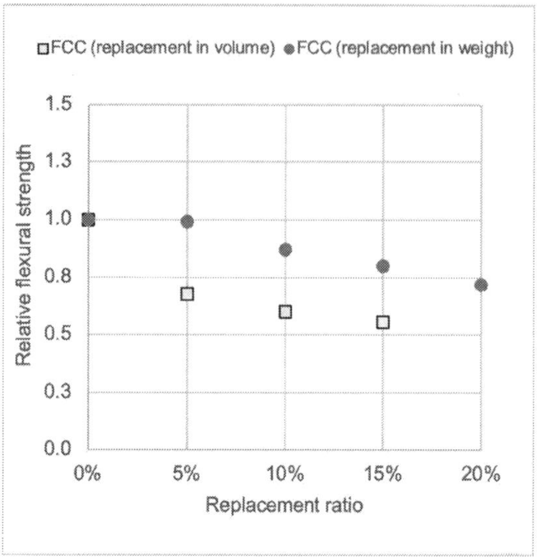

Fig. 5.15 Relative flexural strength between mortars with FCC wastes and the correspondent plain mortar (28 days) [162] [218].

flexural strength was higher with longer fibers and higher incorporation ratios. The mortar with 38.1-mm-long nylon fibers at 1.5% presented an increment of flexural strength of 35%.

However, there are other fiber types that do not increase the mortars' flexural strength, as can easily be noticed in Fig. 5.17. Textile fibers, for example, in some studies increment the mortars flexural strength and in others decrease it. GFRP fibers can also produce increments or decreases. The effectiveness of the fibers is related with the compatibility between the fibers and the matrix, the ratio of incorporation, and the length. Generally, well-designed fibers improve the mortars' flexural strength, due to their "bridging" effect.

5.1.3 Adherence strength

Adherence strength measures the ability of a mortar to adhere to the substrate to prevent its detachment. The lack of adherence is responsible for the detachment of the render mortar.

Every material moves when submitted to humidity or thermal variation conditions. The movement is proportional to their thermal or hydric dilation coefficient, which is a property of each individual material. A render is often submitted to thermal and humidity variations, during a day cycle. In the summer, these mortars can achieve a temperature of up to 65 °C [225] in the hours of greater incidence of solar radiation and overnight decrease to temperatures of 10 °C or less, as in some zones in the South of Europe. The humidity of renders also varies depending on the air humidity.

Table 5.17 Flexural strengths of blended mortars with fibers wastes' addition obtained by other authors (Part 1 of 2).

Type of fiber	Addition percentage	Characteristics	Flexural strength (MPa) 7 days	28 days	90 days	180 days	365 days	References
Sheep's wool	10% (of total volume)	L: 1.5 cm	-	2.72 (+6%)	-	-	-	[X]
		L: 3.0 cm	-	3.12 (+22%)	-	-	-	
	20% (of total volume)	L: 1.5 cm	-	2.53 (–1%)	-	-	-	
		L: 3.0 cm	-	3.56 (+39%)	-	-	-	
Textile	5% (of total volume)	L: 1.5 cm	0.52 (–60%)	1.77 (–16%)	1.77 (+15%)	1.76 (–2%)	1.24 (–36%)	[218]
		L: 3.0 cm	0.87 (–34%)	1.51 (–28%)	1.59 (+3%)	1.76 (–2%)	1.16 (–40%)	
	10% (of total volume)	L: 1.5 cm	-	3.36 (+31%)	3.07 (-5%)	-	-	[223]
		L: 3.0 cm	-	3.48 (+36%)	3.02 (-7%)	-	-	
	20% (of total volume)	L: 1.5 cm	-	2.18 (–15%)	2.22 (-31%)	-	-	
		L: 3.0 cm	-	2.47 (–4%)	2.34 (-28%)	-	-	
	1% (of binder volume)	L: 2 cm	-	1.29 (+10%)	-	-	-	[176]
	4% (of binder volume)	L: 2 cm	-	2.10 (+79%)	-	-	-	
	1% (of binder volume)	L: 4 cm	-	1.33 (+13%)	-	-	-	
	4% (of binder volume)	L: 4 cm	-	2.36 (+102%)	-	-	-	
	0.125% (of binder weight)	-	-	1.29 (+14%)	-	-	-	[178]
	0.25% (of binder weight)	-	-	0.99 (–12%)	-	-	-	
	0.50% (of binder weight)	-	-	1.21 (+7%)	-	-	-	
GFRP	5% (of total volume)	L: 1.5 cm	0.99 (–24%)	2.02 (–4%)	1.76 (+14%)	1.88 (+5%)	1.96 (+1%)	[175]
		L: 3.0 cm	0.97 (–26%)	1.99 (–5%)	1.52 (–13%)	1.58 (–11%)	1.55 (–20%)	
	3% (of sand volume)	Large	6.3 (+23%)	-	-	-	-	[174]
		Medium	4.8 (–5%)	-	-	-	-	
		Small	5.6 (+10%)	-	-	-	-	
		Powder	3.6 (–29%)	-	-	-	-	
	5% (coarse aggregate weight)	EVE1	≈ 2.5 (–4%)	≈ 3.0 (+15%)	-	-	-	[190]
		EVE2	≈ 2.5 (–4%)	≈ 2.7 (+4%)	-	-	-	
		EVE3	≈ 3.0 (+15%)	≈ 3.3 (+27%)	-	-	-	
		UP	≈ 2.6 (Equal)	≈ 3.8 (+46%)	-	-	-	
	0.25% (in volume)	L: 2.0 cm	-	4.9 (+30%)	-	-	-	[202]
	0.75% (in volume)		-	5.4 (+43%)	-	-	-	
	1.25% (in volume)		-	6.0 (+60%)	-	-	-	

Table 5.18 Flexural strengths of blended mortars with fibers wastes' addition obtained by several authors (Part 2 of 2).

Type of fiber	Addition percentage	Characteristics	Flexural strength (MPa)					References
			7 days	28 days	90 days	180 days	365 days	
Acrylic	5% (of total volume)	L: 1.5 cm	1.16 (−11%)	1.78 (−15%)	1.78 (+16%)	1.34 (-25%)	1.86 (-4%)	[218]
		L: 3.0 cm	1.04 (−21%)	1.45 (−31%)	1.52 (−13%)	1.48 (−17%)	1.47 (−24%)	
	2% (of total volume)	L: 12 mm	-	7.8 (+42%)	-	-	-	[224]
Nylon	1% (of cement weight)	L: 12.7 mm	-	5.18 (+16%)	-	-	-	[73]
		L: 25.4 mm	-	5.87 (+32%)	-	-	-	
		L: 38.1 mm	-	5.86 (+32%)	-	-	-	
	1.5% (of cement weight)	L: 12.7 mm	-	5.27 (+18%)	-	-	-	
		L: 25.4 mm	-	5.89 (+32%)	-	-	-	
		L: 38.1 mm	-	6.03 (+35%)	-	-	-	
	1.0% (in volume)	L: 20 mm	-	6.1 (+26.8%)	-	-	-	[74]
		L: 30 mm	-	5.7 (+19.4%)	-	-	-	
		L: 40 mm	-	6.0 (+24.4%)	-	-	-	
	1.5% (in volume)	L: 20 mm	-	6.8 (+41.4%)	-	-	-	
		Straight L: 30 mm	-	5.6 (+17.1%)	-	-	-	
		L: 40 mm	-	6.4 (+34.0%)	-	-	-	
	2.0% (in volume)	L: 20 mm	-	4.8 (−0.1%)	-	-	-	
		L: 30 mm	-	4.3 (−9.8%)	-	-	-	
		L: 40 mm	-	5.2 (+7.3%)	-	-	-	
	0.5% (in volume)	Knotted L: 40 mm	-	5.9 (+22.0%)	-	-	-	
	0.75% (in volume)		-	5.0 (+4.9%)	-	-	-	
	1.0% (in volume)		-	4.5 (−7.4%)	-	-	-	
Broom fibers	0.4% (of cement weight)	L: 20 mm	-	≈ 5.7 (−5%)	-	-	-	[144]
		L: 50 mm	-	≈ 5.8 (−3%)	-	-	-	
	1.8% (of cement weight)	L: 20 mm	-	≈ 5.5 (−8%)	-	-	-	
		L: 50 mm	-	≈ 6.1 (+2%)	-	-	-	

Fig. 5.16 Illustration of a nylon fiber waste from fishing nets activities.

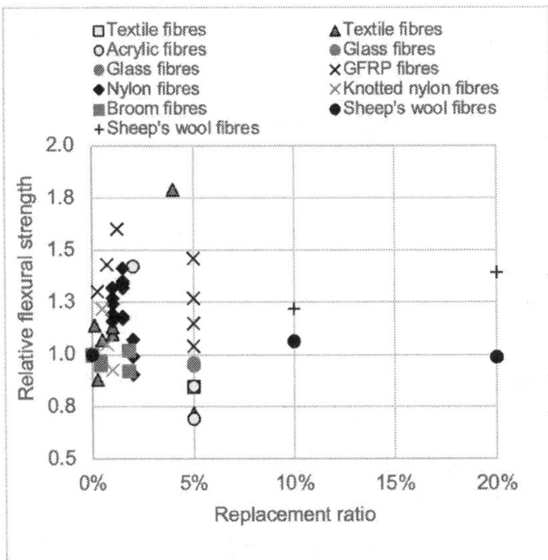

Fig. 5.17 Relative flexural strength between mortars with fiber wastes and the correspondent plain mortar (28 days) [73] [74] [126] [144] [174] [176] [178] [190] [202] [218] [224].

A render can be in contact with an atmosphere with almost 100% of humidity or 40% depending on the day. These actions generate stresses between the render and substrate produced in their interface, which can lead to the detachment of the render in the lower adherence zones. In the nonadherent zones, the thin coat of the render is

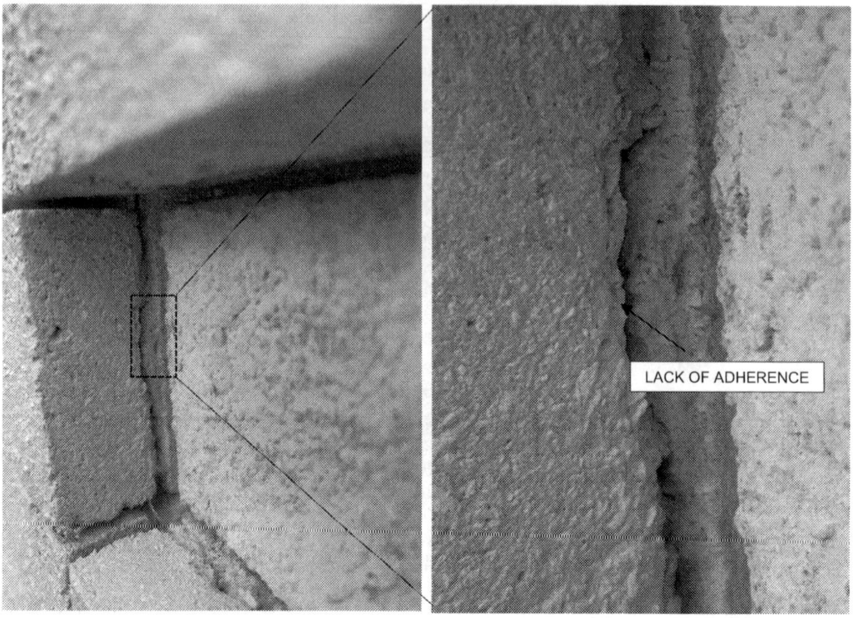

Fig. 5.18 Lack of adherence between a render mortar and the substrate.

submitted to high stresses that are not transmitted to the substrate, and cracking may occur. Additionally, render and substrate, submitted to these gradients of temperatures and humidity, move differently in the detached zones creating high stresses in the zone that it still adherent to the substrate promoting a continuous detachment.

In Fig. 5.18, an example of the lack of adherence between a render mortar and the substrate is represented. This lack of adherence was due to a low compatibility between the mortar and the substrate [218] and resulted in a general microcracking pattern (Fig. 5.19).

Adherence is a characteristic that depends on two materials that are connected. In render mortars, these materials are the mortar and the substrate. The substrate can be a masonry wall, composed of bricks or stone, or a concrete wall, an old render, or other. A porous substrate is favorable to promote the adherence strength between the substrate and the render. It will absorb part of the binder's slurry allowing the mixture to penetrate the substrate capillaries, improving the bond between these two materials. The characteristics of the mortars also have influence on the mortar/substrate bond. Workability is a characteristic that influences the bond between the render and the substrate. A mortar with low workability, namely with low water content, is unable to adhere to the substrate as it cannot completely fill the rugosity of the substrate surface and it does not have enough water to penetrate into the substrate pores. Water retention is another property that can influence the mortar/substrate's bond. A high water retentivity of the mortar avoids quick desiccation of the hydraulic binder, allowing the hydraulic reactions to take place, which improves both adherence and the mortar's

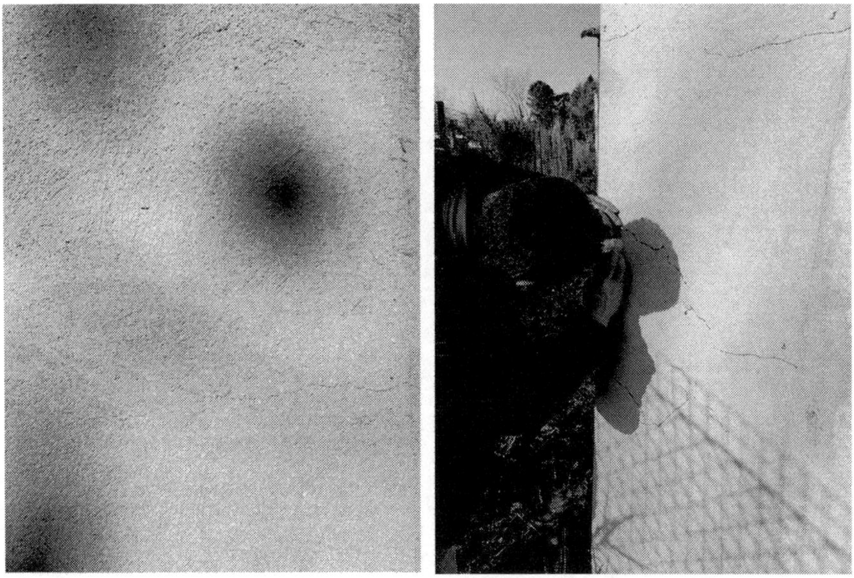

Fig. 5.19 Cracking due to lack of adherence between render mortar and substrate.

mechanical strength. Nevertheless, if a mortar retains an excessive volume of water, it does not allow the substrate to adequately absorb the mortar slurry, which can be detrimental to adherence.

Adherence strength can be evaluated by a pull-off test. In this test, as described in Section 2.3.4, a core that cuts through all the layers of the render is cut, until the substrate is reached. Then, a direct load is applied perpendicularly to the surface of the specimen, pulling the core. The mortars are tensioned, which also creates tensions on the mortar/substrate interface's bond. The fracture can occur in the interface, in the mortar, or in the substrate. Thus, the adherence test's fracture can be adhesive or cohesive. Adhesive fracture is the one that is produced in the interface between the mortar and the substrate. Cohesive is a fracture that occurs in the mortar or in the substrate. Only in adhesive fracture does the value obtained in the test represent the exact value of the adherence strength. In cohesive fracture, the value obtained is the maximum tensile strength of the material in which the fracture occurred and the adherence strength is known to be higher than that value.

Wastes incorporation modifies the mortars' properties and consequently can also modify the mortars' adherence strength. Table 5.19 presents some examples of adherence strength between the mortars and a ceramic surface and Fig. 5.20 illustrates the relative adherence between the modified mortars and their reference mortar.

Some aggregate wastes such as red ceramics, sanitary ware, or construction and demolition waste (mainly concrete) presented adherence strengths higher than the reference mortar. In red ceramics and sanitary ware, both replacing the natural aggregate at 50%, the adherence fracture was adhesive in the interface. When sanitary ware

Table 5.19 Adherence strength and fracture type of mortars with wastes' incorporation from several authors.

Type of incorporation	Waste	0%	5%	10%	15%	20%	50%	100%	References
Aggregates	Red ceramics	0.34	-	-	-	-	0.40	-	[221]
		B					A		
	Sanitary ware	0.63	-	-	-	-	0.67	-	[18]
		A					A		
		0.47	-	-	-	-	-	0.50	[218]
		B						B	
	Concrete	0.33	-	-	-	0.27	-	-	[8]
		B				B			
	CDW (predominantly concrete)	0.54	-	-	-	-	-	0.90	[155]
		A/B						A/B	
	CDW	0.54	-	-	-	0.48	-	-	[155]
		A/B				A/B			
	Glass	0.72	-	-	-	0.65	0.49	0.62	[143]
		B				B	B	B	
	Shredded tires	0.51	-	-	0.34	-	-	-	[148]
		A/B			A/B				
	PET flakes	0.33	-	-	0.21	-	-	-	[137]
		-			-				
	PET pellets	0.33	-	-	0.26	-	-	-	[137]
		-			-				
Filler	Red ceramics powder	0.34	-	0.43	-	-	-	-	[49]
		B		A					
	Sanitary ware powder	0.63	-	-	-	0.71	-	-	[7]
		A				A			
	Concrete powder	0.33	-	-	0.45	-	-	-	[51]
		A/B			A				
	CDW (predominantly concrete) powder	0.54	-	-	-	0.87	-	-	[154]
		A/B				A			
	CDW powder	0.54	-	-	0.66	-	-	-	[108]
		B			A				
	Glass powder	0.64	-	-	-	0.67	-	-	[142]
		B				A/B			
	GFRP powder	0.47	-	-	-	-	0.86	-	[175]
		B					C		
Binder	Forest biomass ashes	0.47	-	-	0.26	-	-	-	[175]
		B			B				
Fibre	Textile fibres	0.47	0.44	-	-	-	-	-	[175]
		B	B						

Note: Fracture pattern A, Adhesive fracture (in the interface mortar/substrate); Fracture pattern B, Cohesive fracture (in the mortar); Fracture pattern C, Cohesive fracture (in the substrate).

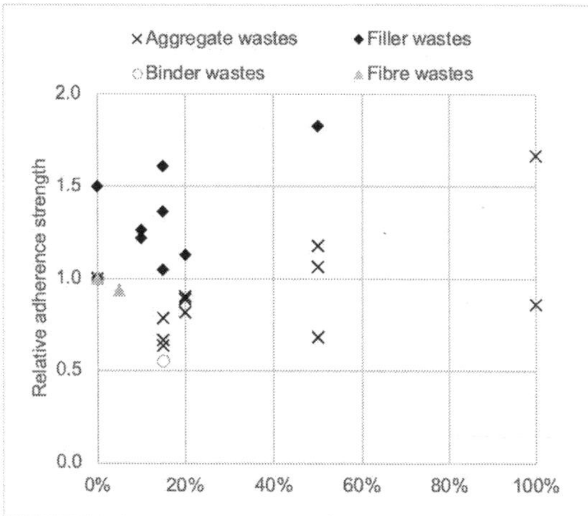

Fig. 5.20 Relative adherence strength of mortars with aggregate, filler, binder, or fiber wastes [7] [8] [18] [49] [51] [137] [142] [143] [148] [154] [155] [218] [221].

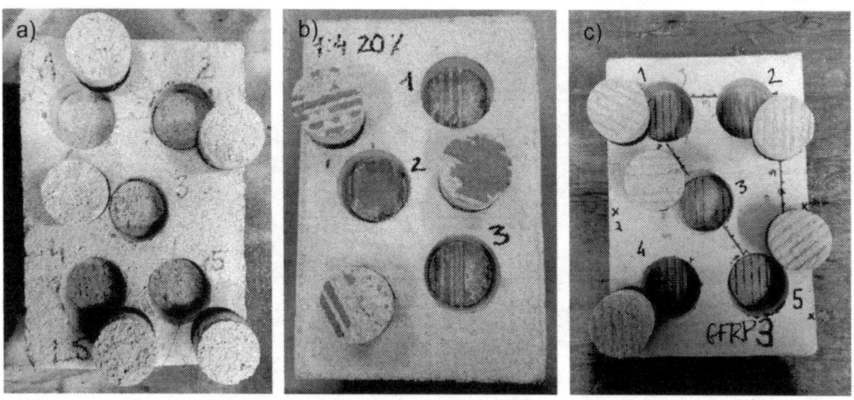

Fig. 5.21 Mortar adherence strength pattern illustrations. (A) Sanitary ware waste (replacing natural aggregate at 100%). (B) Sanitary ware powder waste (replacing natural aggregate at 20%). (C) Glass fiber–reinforced polymer (replacing natural aggregate at 50%).

replaced natural aggregate at 100% the fracture was cohesive in the mortar (illustrated in Fig. 5.21), similar to CDW waste (predominantly concrete). However, the improvement of adherence strength when aggregate wastes are incorporated may not be extrapolated to all wastes, as noticed in Fig. 5.20. Concrete waste or construction and demolition waste, replacing natural aggregate at 20%, reduced the adherence strength by about 11%–18%. The incorporation of glass waste replacing natural aggregate, as investigated by Penacho et al. [143], also resulted in lower adherence strengths. Polymer wastes, such as shredded tires, PET flakes, or PET pellets, replacing natural aggregate reduced the adherence strength value as well.

Filler wastes, on the other hand, present, in general, higher values in adherence strength tests. Contrary to aggregate wastes, these wastes tend to present a fracture type A or C, meaning they present fractures that are adhesive in the interface or cohesive in the substrate. Thus, it can be concluded that the flexural strength of mortars with filler wastes does not influence the test value. In fact, the mortars' tensile strength is higher than the interface's strength and therefore the mortars' tensile strength does not interfere with the adherence strength test. Fig. 5.21 illustrates the fracture pattern of a mortar with 20% of sanitary ware powder and a mortar with 50% of a GFRP powder, both presenting in general an adhesive fracture pattern.

In Table 5.19, an example of adherence strength of a mortar incorporating a binder waste is presented. The results refer to a forest biomass ash waste that replaced cement at 15% (in volume). The waste was responsible for reducing the adherence strength by 44%. In this case, the lower tensile strength of the modified mortar is the reason for a high reduction of the adherence strength, which presents a cohesive fracture in the mortar. A reduction of mechanical strength is a trend of the binder wastes. These wastes do not present the hydraulic strength capability of cement and therefore a worse mechanical behavior must be expected. However, the binder waste example (forest biomass ashes) presenting a fracture type B (Fig. 5.22) means that the adherence strength at the mortar/substrate interface is higher than the value obtained by the research [218]. Notwithstanding the reduction of adherence strength, the value

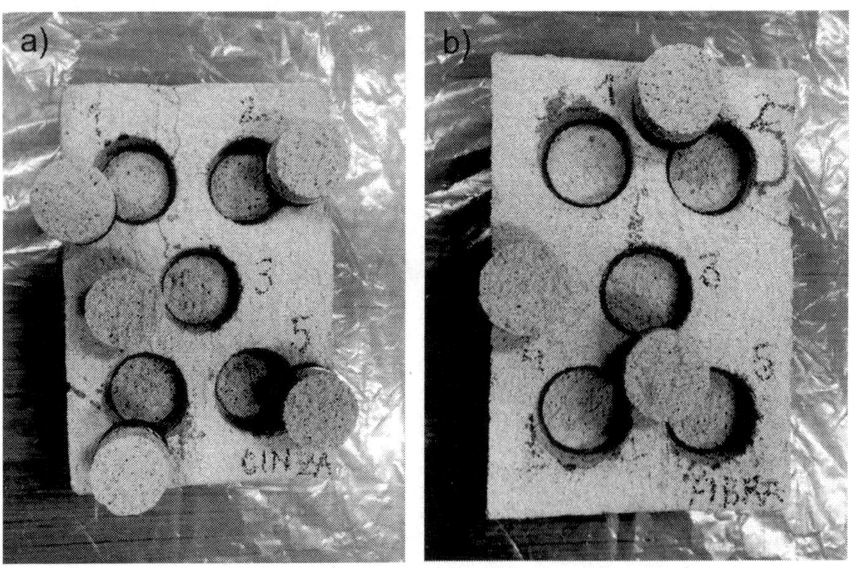

Fig. 5.22 Mortar adherence strength pattern illustrations. (A) Forest biomass ash waste (replacing cement by 15%). (B) Textile fibers addition (5% of total volume).

obtained was reasonable and, according to the relevant National Report from LNEC [247], the adherence strength of the mortar meets the minimum requirement to be applied as a render (higher than 0.30 MPa or cohesive fracture). Gonilho-Pereira et al. [178] also evaluated the adherence strength of mortars with textile fibers and found a decrease for all modified mortars.

The adherence strengths obtained were 0.45, 0.27, and 0.35 MPa, respectively for additions of 0.125%, 0.25%, and 0.50%, which corresponded to a decrease in comparison with the reference mortar of 13%, 47%, and 32%.

An example of a fiber waste is also presented in Table 5.19. This example is of a 1.5 cm long textile fiber waste that was added to reinforce a conventional mortar. This mortar required a high water/binder ratio that increased the matrix porosity; however, it presented an adherence strength value similar to that of the reference mortar: 0.44 MPa versus 0.47 MPa [218]. This mortar, similar to its reference counterpart, presented a fracture type B, meaning that both presented a fracture in the mortar. These results are in agreement with the flexural strength tests. In fact, the mortars presented a similar flexural strength and, as the fracture was cohesive in the mortars, they presented similar adherence strength values.

As noticed, most of the wastes and reference mortars presented a fracture type B, cohesive in the mortar, confirming that the mortars' tensile strength is the property that conditions the adherence test results. Fig. 5.23 illustrates the correlation between adherence strength and the maximum stress from the flexural strength test. A clear trend between these two properties can be found. In fact, the higher the flexural strength a mortar has, the higher the value of adherence strength gets. However, it must be highlighted that a fracture type B means that the strength in the interface mortar/substrate is higher than the value obtained in the adherence strength test.

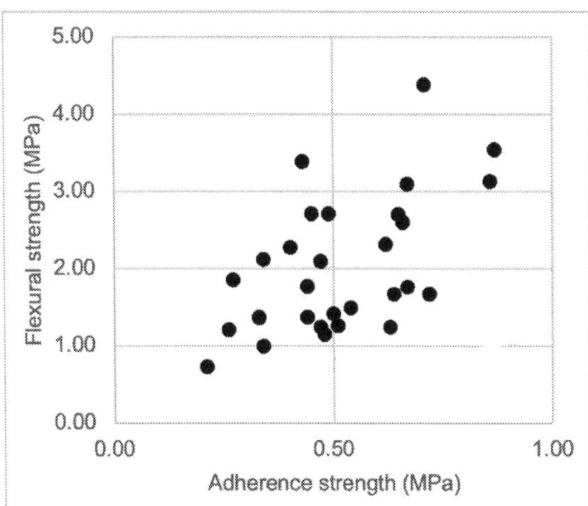

Fig. 5.23 Flexural versus adherence strength [7] [8] [18] [49] [51] [137] [142] [143] [148] [154] [155] [218] [221].

5.1.4 Impact resistance

Renders and plasters should be resistant to impacts. In render mortars, these impacts include small blows from people or objects, namely balls. Plasters also must be resistant to small impacts that result from regular householders' actions but may also include specific types of use. Gymnasiums or hospitals are specific utilization types in which blows are stronger and it is not expected that plasters should crack under those actions.

Impact resistance, as mentioned in Section 2.3.6., can be measured by a pendulum or, in laboratory, by the fall of a mass from different heights. The height that originates the first crack, multiplied by the falling mass weight and by the gravity acceleration, gives the impact energy that the mortar is no longer able to absorb. In addition to the height that corresponds to the first crack, the cracking width, diameter and depth of the dent produced by the falling mass can be analyzed and give an indication about the deformability of the render. A more deformable mortar absorbs higher impact energy and achieves a more ductile pattern with the appearance of the first crack: larger dent diameters, higher dent depths, and lower cracking widths. The test is performed until failure, and for that the falling mass height is increased until failure occurs. The impact energy in failure is calculated as in the first crack (multiplying the height by the falling mass weight and by the gravity acceleration). More deformable mortars usually present a higher number of cracks although more distributed in the specimens with lower widths. Brittle mortars usually present single cracks with larger widths.

The resistance to impact results is related with the deformability and compressive strength of the mortars [218]. The energy absorbed by the renders until the first crack occurs depends on the deformability of the mortars, as well as on the failure pattern, the more deformable mortars presenting more ductile patterns. The impact energy for that failure to occur is more dependent on the mortar's compressive strength. As discussed in this chapter, wastes' incorporation in mortars influences some of the mortars' properties, where deformability and compressive strength are also included. These modifications also influence the mortars' impact resistance.

Some examples of the impact energy necessary for the first crack to occur, in mortars with wastes' incorporation, are presented in Table 5.20. The relative impact energy between modified mortars and the respective reference mortar is also illustrated in Fig. 5.24. Four aggregate wastes, a binder waste, and a fiber waste are presented. The aggregates include sanitary ware (replacing natural aggregate at 100%),

Table 5.20 Impact energy (J) for the first crack to occurs.

Type of incorporation	Waste	0%	5%	10%	15%	100%	References
Aggregates	Sanitary ware	10.78	-	-	-	10.78	[218]
	Shredded tires	10.78	10.78	12.74	14.70	-	[148]
	Plastics (PET flakes)	11.76	12.25	14.41	14.01	-	[137]
	Plastics (PET pellets)	11.76	11.76	13.43	12.05	-	[137]
Binder	Forest biomass ashes	10.78	-	-	11.76	-	[218]
Fibre	Textile fibres (1.5 cm long)	10.78	12.74	-	-	-	[218]

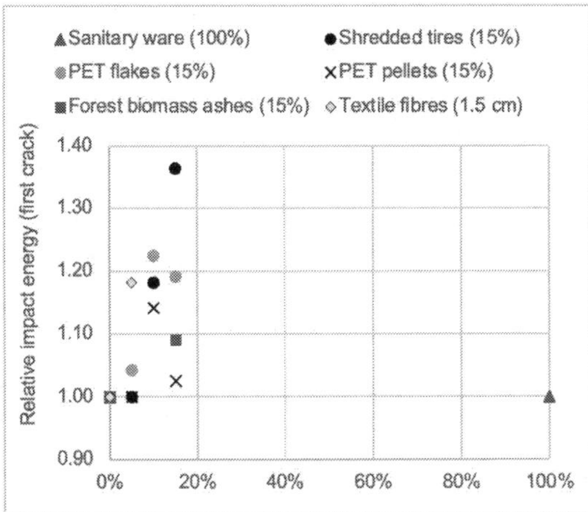

Fig. 5.24 Relative impact resistance in first crack between reference mortar and modified mortars by wastes' incorporation [137] [148] [218].

shredded tires (replacing natural aggregate at 5%, 10%, and 15%), and two PET wastes, PET flakes and PET pellets, both replacing the natural aggregate at 5%, 10%, and 15%. The binder waste presented as an example is forest biomass ashes, which replaced the binder at 15%. Finally, textile fibers were the fiber waste considered.

In all the examples shown, the wastes presented an equal or higher impact energy until the first crack occurred than the respective reference mortar. These results mean that the mortars presented are more deformable than the reference mortar. The shredded tires and the PETs are polymers, presenting a lower modulus of elasticity than the natural aggregate, and therefore an improvement of impact resistance was expected. Shredded tires were able to improve the impact energy by 36%. The improvement of impact resistance is also a characteristic of fibers' addition. Textile fibers, which correspond to a waste from the textile industry, also improved the impact energy until the first crack by almost 20%. For other types of wastes, a beneficial improvement of impact resistance may not be clearly evident as for sanitary ware or forest biomass ashes, which also improved this property of the mortars. It is necessary to refer that although all the wastes analyzed as examples improved the impact energy in the first crack, the improvement cannot be generalized for all wastes. Fillers, for instance, significantly increase the hardness of the mortars, which may result in more brittle mortars, as discussed in Section 5.2.1, possibly resulting in worse behavior concerning impact actions.

The characteristics of the dent caused by the impact of the falling mass in terms of diameter and depth were analyzed by Farinha [218], when researching the influence of several wastes in mortars' impact resistance (Table 5.21). It was noticed that although sanitary ware mortars presented a similar energy impact for the first

Table 5.21 Dent and cracking characteristics by the impact resistance test.

Waste	Characteristics	0%	5%	15%	100%	References
Sanitary ware	Dent diameter (mm)	11.99	-	-	14.57	[218]
	Dent depth (mm)	1.25	-	-	1.73	
	Crack width (mm)	0.40	-	-	0.05	
Forest biomass ashes	Dent diameter (mm)	11.99	-	14.90	-	[218]
	Dent depth (mm)	1.25	-	1.63	-	
	Crack width (mm)	0.40	-	0.10	-	
Textile fibres (1.5 cm long)	Dent diameter (mm)	11.99	15.83	-	-	[218]
	Dent depth (mm)	1.25	1.63	-	-	
	Crack width (mm)	0.40	0.10	-	-	

crack to occur, the dent characteristics and the cracking patterns indicated a more ductile behavior than that of the reference mortar. The diameter and depth of the dent were larger than the ones presented by the reference mortar and the crack obtained presented a significant lower width (0.05 mm vs 0.40 mm). The results can also be observed comparing Figs. 5.25 and 5.26. The crack in sanitary ware is quite difficult to observe with the naked eye, while the one in the reference mortar is visible. Forest biomass ashes mortar, an example of a binder waste, also presented larger diameters and dents depths and lower crack widths (0.10 mm vs 0.40 mm). The lower modulus of elasticity of this mortar justifies a better behavior to impact. In Fig. 5.27, the cracking pattern of this mortar with incorporation of forest biomass ashes is presented.

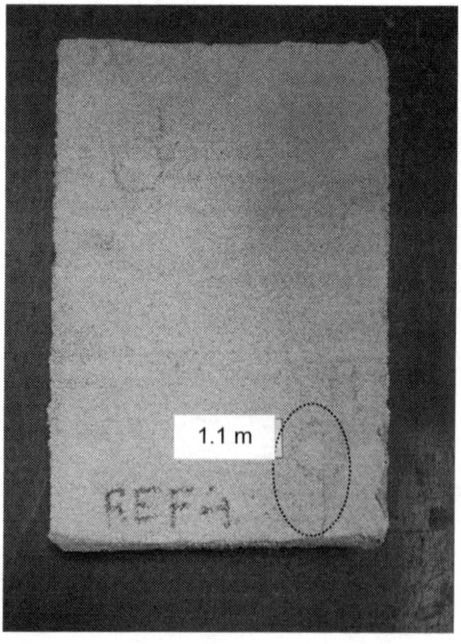

Fig. 5.25 First crack pattern in the impact resistance test of a reference mortar [218].

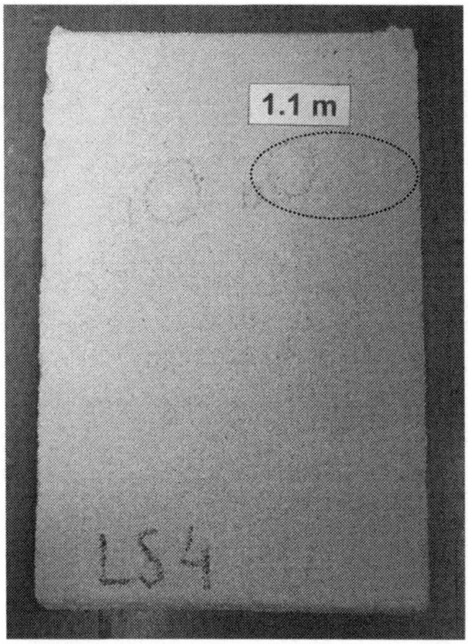

Fig. 5.26 First crack pattern in the impact resistance test of a mortar with sanitary ware wastes as aggregate [218].

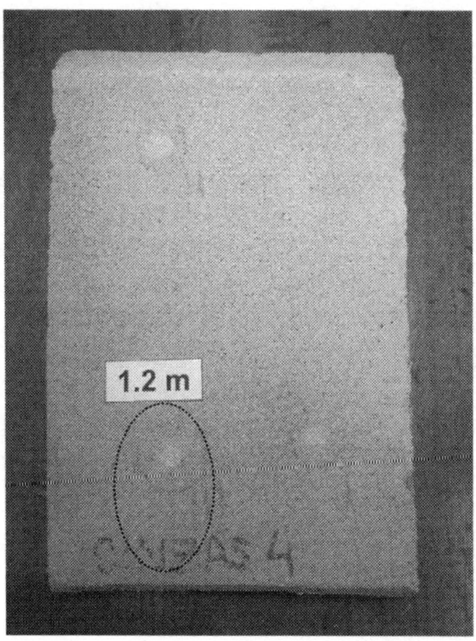

Fig. 5.27 First crack pattern in the impact resistance test of a mortar with forest biomass ashes incorporation. [218].

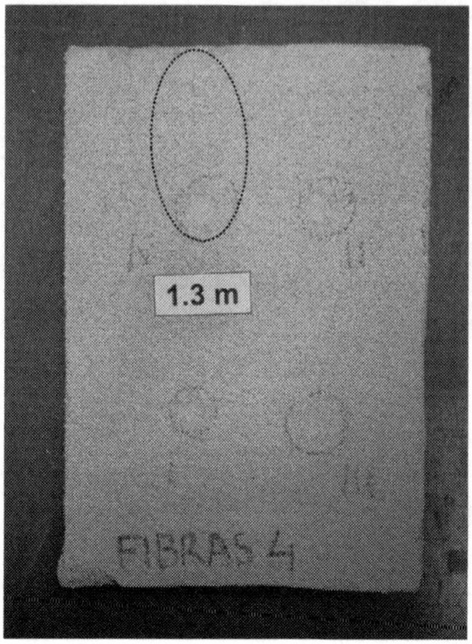

Fig. 5.28 First crack pattern in the impact resistance test of a mortar reinforced with textile fibers (1.5 cm long) [218].

Farinha [218] also investigated the dent and cracking characteristics of a mortar reinforced with textile fibers waste up to the appearance of the first crack in the impact test (Fig. 5.28). Similar to the previous ones, a higher dent diameter and depth of the mortars was noticed in comparison with the reference mortar. The crack width was also lower than that of the reference mortar, which also indicates that the fibers improved the mortars' deformability.

The resistance to impact until the first crack occurs is essential to render mortars or to plasters that can be submitted to blows, in hospitals, gymnasium, or others, as mentioned. The examples of wastes' incorporation improved this characteristic of the mortars. However, other wastes, namely the ones that confer more hardness to the mortars, such as filler wastes, may not improve this characteristic. The wastes presented previously have a lower modulus of elasticity when compared with a reference mortar and are therefore responsible for more ductile behaviors than that of the conventional mortars.

Mortars with or without wastes were evaluated by some authors until failure was achieved [137] [218]. The impact energy and the cracking pattern are evaluated. In Table 5.22, the impact energy in failure of mortars with wastes' addition is summarized and, in Fig. 5.29, the relative impact between the modified mortars and the reference mortar is illustrated.

Sanitary ware (aggregate waste), forest biomass ashes (binder waste), and PET pellets (5%) presented an impact energy in failure 7% lower than the reference mortar. The remaining modified mortars presented higher impact energy than the reference mortar.

Table 5.22 Impact energy (J) in failure.

Type of incorporation	Waste	0%	5%	10%	15%	100%	References
Aggregates	Sanitary ware	13.72	-	-	-	12.74	[218]
	Plastics (PET flakes)	13.72	14.01	16.66	16.37	-	[137]
	Plastics (PET pellets)	13.72	12.74	14.70	14.01	-	[137]
Binder	Forest biomass ashes	13.72	-	-	12.74	-	[218]
Fibre	Textile fibres (1.5 cm long)	13.72	14.70	-	-	-	[218]

Impact energy, until first crack or until failure, is correlated with mortars' compressive strength and modulus of elasticity. On the one hand, the mortar's deformability allows it to absorb part of the energy produced in the impact, a mortar more deformable being able to absorb higher energy than a less deformable one. Fig. 5.30 represents the correlation between the modulus of elasticity and both the impact energies analyzed (until first crack and until failure). It can be noticed that, in general, higher impact energies occur for mortars with lower modulus of elasticity. On the other hand, mortars must withstand the impact, that is have an adequate compressive strength, to maintain their integrity. However, high compressive strength is related to a harder mortar surface and thus to more brittle mortar and consequently with lower

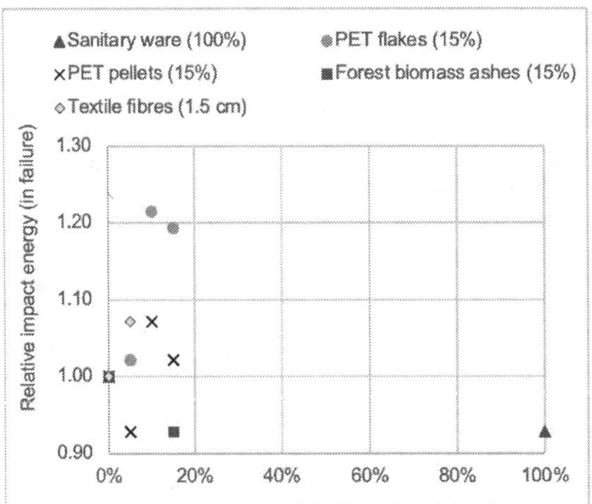

Fig. 5.29 Relative impact resistance in failure between reference mortar and modified mortars by wastes' incorporation [137] [218].

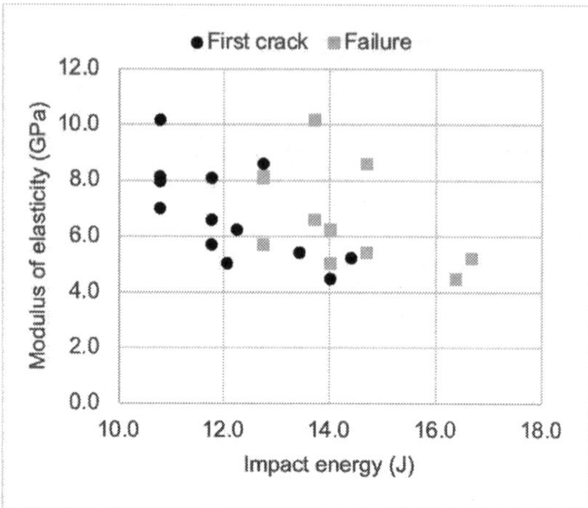

Fig. 5.30 Correlation between modulus of elasticity and impact resistance [137] [148] [218].

impact energy, as can be seen in Fig. 5.31. In fact, compressive strength and modulus of elasticity are related to the impact energy absorbed by the mortars.

In addition to the values obtained in the impact resistance test, the failure pattern also gives information about the mortar's brittleness or ductility. The failure pattern of the mortars tested and presented previously is illustrated in Figs. 5.32–5.35. The sanitary ware waste mortar and the reference mortar achieved failure producing a single crack, which crossed entire specimen and was responsible for the substrate's damage. The width of the crack was larger for the reference mortar; however, in both cases, the width was larger than 2 mm. These failure patterns are typical characteristics of

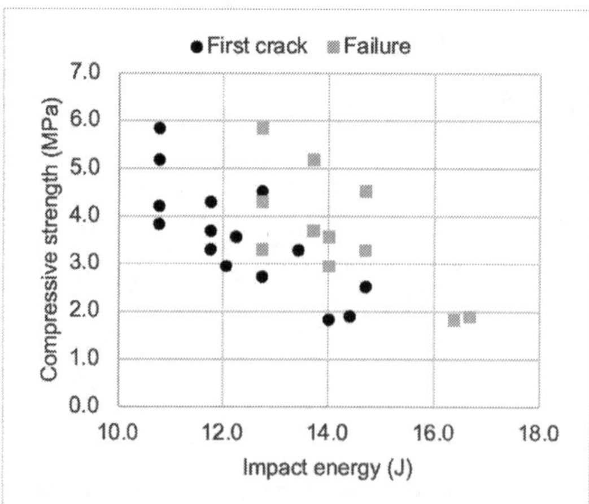

Fig. 5.31 Correlation between compressive strength and impact resistance [137] [148] [218].

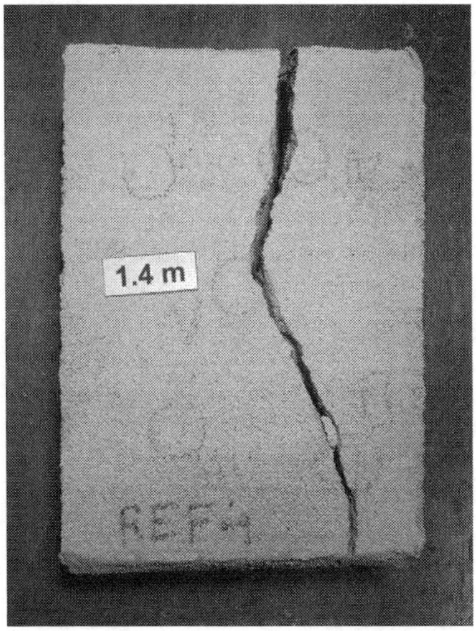

Fig. 5.32 Failure pattern in the impact resistance test of a reference mortar [218].

a brittle material. The mortar with forest biomass ashes and the mortar with fibers addition presented several distributed cracks all over the specimen, indicating a more ductile pattern. In the fibers example, it can be noticed that, in fact, the substrate achieved failure simultaneously with the mortar.

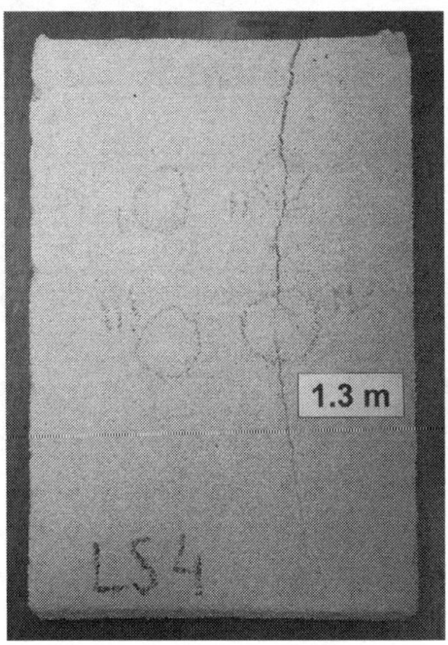

Fig. 5.33 Failure pattern in the impact resistance test of a mortar with sanitary ware wastes as aggregate [218].

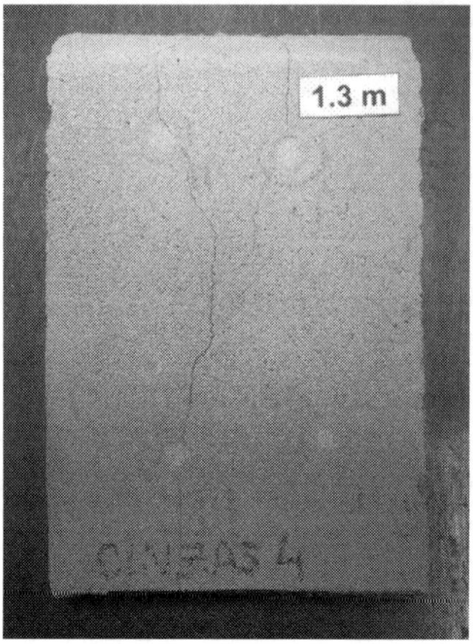

Fig. 5.34 Failure pattern in the impact resistance test of a mortar with forest biomass ashes incorporation [218].

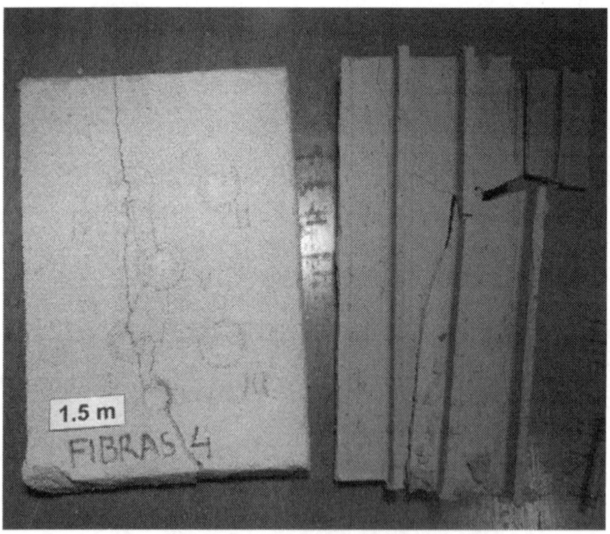

Fig. 5.35 Failure pattern in the impact resistance test of a mortar reinforced with textile fibers (1.5 cm long) [218].

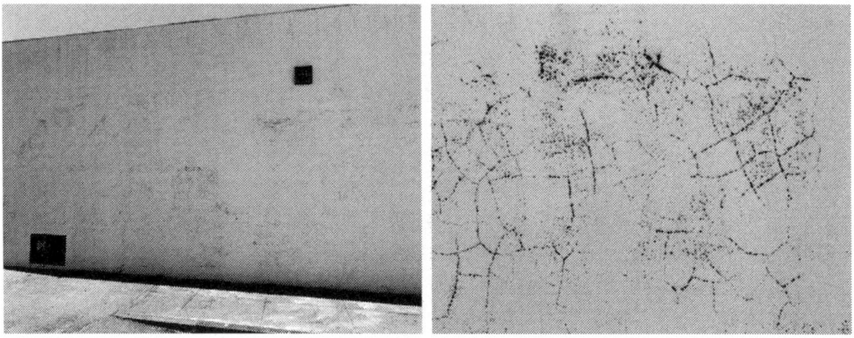

Fig. 5.36 Illustration of cracking due to shrinkage and the effect of this cracking on biocolonization.

5.2 Cracking

Render mortars' cracking can occur in the fresh or in hardened state. Usually, in the fresh state, they crack due to plastic shrinkage or to lack to adherence to the substrate. In the hardened state, mortars can crack due to several actions, such as shrinkage (Fig. 5.36), thermal variation, humidity variation, ice formation, movements of the substrate, lack of adherence, blows, or chemical reactions that involve the expansion or contraction of the mortars. The deformability of the render mortars, and in general their ability to absorb energy, can minimize the cracking when submitted to the mentioned actions. This section is dedicated to the mortars' deformability and toughness (Fig. 5.37).

Fig. 5.37 Determination of the modulus of elasticity of a render mortar applied over a ceramic brick wall by the method of ultrasound pulse velocity.

5.2.1 Elastic deformability

In addition to mechanical strength, it is necessary to evaluate the mortars' deformability to analyze their mechanical behavior. Some degree of deformability is necessary to accommodate stresses and prevent cracking.

Deformability is a characteristic of each material and is related to its ability to expand or contract. For most materials, mortars included, there is an elastic deformation and a plastic deformation before failure occurs. During the phase of plastic deformation, if the load is removed, the material is no longer able to reach the same shape that it had before the deformation. Render mortars should be able to present a relatively high elastic deformability to accommodate the stresses that they are submitted to, while maintaining the same characteristics after the solicitations. The modulus of elasticity allows evaluating the elastic deformability.

The incorporation of aggregate wastes, replacing natural ones, modifies the mortars' microstructure and consequently their deformability. In Table 5.23, some examples of the modulus of elasticity of render mortars with aggregate wastes are summarized. The authors of the studies presented as examples studied render mortars with a volumetric ratio of 1:4 and the binder used was CEM II/B-L 32.5 N cement. The modulus of elasticity was analyzed by the frequency of resonance. In Fig. 5.38, the relative modulus of elasticity between the modified mortars and their correspondent reference mortar is also presented.

It can be noticed that, in general, aggregate wastes decrease the modulus of elasticity of the mortars, when compared with the corresponding reference mortars. This reduction means that these wastes improve the deformability of the mortars. Render mortars with aggregate wastes' incorporation result in more deformable mortars. This decrease in modulus of elasticity is mainly related to the porosity of the mortars. Aggregate wastes, due to their more angular shape, do not rearrange so efficiently as the natural aggregate, usually characterized by spherical particles. This angularity hinders the rearrangement of the particles, creating more empty voids between them. In addition, aggregate wastes, also due to their shape, increment the mortars' water needs, as discussed in Section 4.2. The water needed in the renders to achieve a suitable workability is not completely combined in the hydraulic reactions of the binder, giving rise to voids after drying. This higher porosity was responsible for a decrease of the modulus of elasticity.

Usually, the modulus of elasticity is related to the mortars' compressive strength. In mortars with aggregate wastes, however, the modulus of elasticity obtained by the frequency of resonance is lowered by the incorporation of some wastes that presented higher compressive strength. For example, the incorporation of sanitary ware and of concrete wastes for 100% replacement of aggregates resulted in mortars with a compressive strength higher than the corresponding reference mortar's by more than 20% and, simultaneously, presented a modulus of elasticity 20% lower than that of the reference mortar, thus being, theoretically and simultaneously, more deformable and stronger.

Concerning filler wastes, they are characterized for incrementing the mortars' compactness, which has been discussed in this book. This compactness is responsible

Table 5.23 Modulus of elasticity of blended mortars with aggregate wastes' incorporation obtained by several authors.

Type of fiber	Incorporation ratio	Modulus of elasticity (GPa)				References
		28 days	60 days	90 days	150 days	
Red ceramics	50%	-	8.7 (-40%)	-	8.3 (-40%)	[221]
Sanitary ware	20%	11.9 (+1%)	-	11.5 (+3%)	-	[18]
	50%	11.6 (−1%)	-	11.1 (−1%)	-	
	100%	8.2 (−30%)	-	8.1 (−28%)	-	
Concrete	20%	8.3 (+17%)	-	-	-	[8]
CDW (predominantly concrete)	20%	10.2 (−7%)	-	8.9 (−25%)	-	[155]
	50%	8.9 (−9%)	-	7.4 (−29%)	-	
	100%	8.3 (−25%)	-	6.1 (−42%)	-	
CDW	20%	10.8 (−2%)	-	9.8 (−6%)	-	[109]
	50%	6.4 (−42%)	-	5.8 (−44%)	-	
	100%	4.5 (−59%)	-	4.0 (−62%)	-	
Glass	20%	10.2 (+41%)	-	9.3 (+48%)	-	[143]
	50%	11.0 (+53%)	-	10.0 (+58%)	-	
	100%	9.0 (+24%)	-	8.3 (+32%)	-	
Shredded tires	5%	7.0 (−18%)	-	6.8 (−20%)	-	[148]
	10%	5.7 (−28%)	-	5.7 (−33%)	-	
	15%	5.4 (−52%)	-	4.8 (−44%)	-	
PET flakes	5%	6.3 (−5%)	-	5.8 (−8%)	-	[137]
	10%	5.2 (−21%)	-	4.3 (−32%)	-	
	15%	4.5 (−32%)	-	4.1 (−36%)	-	
PET pellets	5%	5.7 (−13%)	-	5.3 (−17%)	-	[137]
	10%	5.4 (−18%)	-	4.7 (−26%)	-	
	15%	5.0 (−24%)	-	4.6 (−28%)	-	

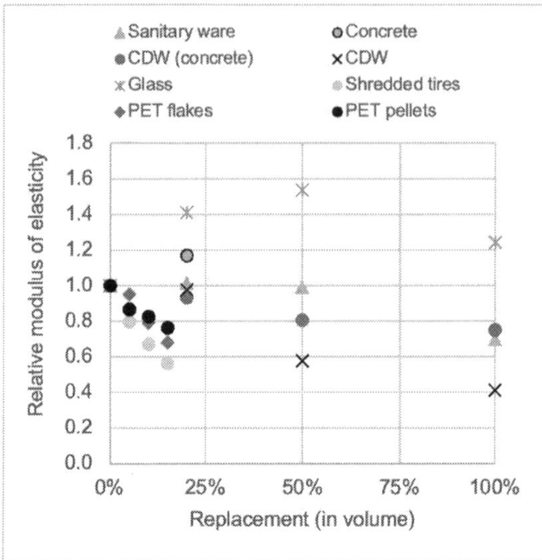

Fig. 5.38 Relative modulus of elasticity between mortars with aggregate wastes and the corresponding plain mortar (28 days) [8] [18] [137] [143] [148] [155] [221].

for a considerable increment of mechanical strength and modulus of elasticity. In Table 5.24, some examples of filler wastes in render mortars and their influence on the modulus of elasticity are presented. The relative modulus of elasticity between the modified mortars and their corresponding plain mortar is illustrated in Fig. 5.39.

From Fig. 5.39, it is noticed that only in two examples the filler wastes presented an equal modulus of elasticity to that of their reference mortar. The polymeric waste considered as an example (GFRP) in filler wastes presented a similar modulus of elasticity to that of the reference mortar for all ratios of incorporation [176]. Tittarelli and Shah [65], who incorporated a similar waste, found a decrease by 20% and 50% for 5% and 10% GFRP content. CDW incorporated as a filler incremented the modulus of elasticity by 20% when incorporated at 10%. However, with the increment of filler content (15% and 20%), the modulus of elasticity was reduced and became similar to the reference mortar's. The remaining filler wastes presented in Fig. 5.39 considerably incremented the modulus of elasticity. Glass waste, for example, when incorporated as an aggregate at 20%, had a modulus of elasticity (at 28 days) of 10.2 MPa [143]; however, when incorporated as a filler in the same percentage (20%), presented a modulus of elasticity of 12.5 MPa [142], which corresponded to an increment compared to the reference mortar of about 75%. These high increments of the modulus of elasticity indicate that filler wastes make the mortars less deformable. In structural mortars, for example bedding mortars, this could not be limitative, and even be an advantage, but in render mortars this high reduction of deformability can be responsible for mortars' microcracking. Although this type of mortars presents high strengths, they may not be able to deform conveniently to accommodate the stresses induced in renders during their life cycle.

Table 5.24 Modulus of elasticity of blended mortars with filler wastes' incorporation obtained by several authors.

Type of fiber	Incorporation ratio	Modulus of elasticity (GPa)							References
		28 days	60 days	90 days	150 days	180 days	365 days	730 days	
Red ceramics	10%	-	12.2 (-17%)	-	15.6 (+12%)	-	-	-	[49]
Sanitary ware	10%	15.7 (+34%)	-	14.6 (+30%)	-	-	-	-	[7]
	15%	17.4 (+49%)	-	16.1 (+44%)	-	-	-	-	
	20%	18.6 (+59%)	-	17.3 (+54%)	-	-	-	-	
Concrete	15%	10.6 (+49%)	-	-	-	-	-	-	[51]
CDW (predominantly concrete)	10%	14.1 (+28%)	-	12.2 (+17%)	-	-	-	-	[154]
	15%	15.4 (+40%)	-	14.3 (+38%)	-	-	-	-	
	20%	15.2 (+38%)	-	15.3 (+47%)	-	-	-	-	
CDW	10%	13.3 (+21%)	-	12.0 (+15%)	-	-	-	-	[154]
	15%	11.9 (+8%)	-	11.7 (+13%)	-	-	-	-	
	20%	10.6 (+4%)	-	9.8 (−6%)	-	-	-	-	
Glass	10%	9.9 (+37%)	-	8.7 (+39%)	-	-	-	-	[142]
	15%	11.0 (+52%)	-	10.1 (+60%)	-	-	-	-	
	20%	12.5 (+74%)	-	10.9 (+73%)	-	-	-	-	
GFRP	10%	9.6 (−6%)	-	9.2 (+15%)	-	9.6 (+10%)	9.3 (-4%)	9.4 (−7%)	[172]
	15%	9.7 (−5%)	-	10.2 (+28%)	-	10.4 (+20%)	10.1 (+4%)	9.9 (−2%)	
	20%	10.2 (Equal)	-	11.1 (+39%)	-	11.1 (+28%)	9.7 (Equal)	11.7 (+16%)	
	50%	9.9 (−3%)	-	10.0 (+25%)	-	10.2 (+17%)	10.8 (+11%)	11.0 (+9%)	

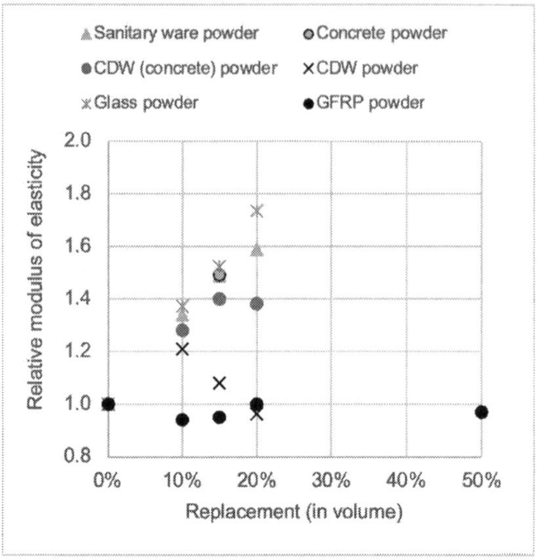

Fig. 5.39 Relative modulus of elasticity between mortars with filler wastes and their plain mortar (28 days) [7] [49] [51] [142] [154] [172].

Concerning the wastes that replaced the binders, designated in this book as binder wastes, two examples of how these wastes can affect the modulus of elasticity when incorporated in mortars are given in Table 5.25 and in Fig. 5.40. The examples presented are from render mortars with a volumetric ratio of 1:4 (CEM II/B-L 32.5 N:

Table 5.25 Modulus of elasticity of blended mortars with binder wastes' incorporation obtained by other authors.

Wastes	Incorporation ratio	Modulus of elasticity (GPa)					References
		28 days	90 days	180 days	365 days	730 days	
Forest biomass ashes	5%	9.9 (−3%)	9.0 (+13%)	7.6 (−13%)	7.6 (−22%)	7.7 (−24%)	[215]
	10%	9.4 (−8%)	8.2 (+3%)	7.1 (−18%)	6.8 (−30%)	6.9 (−32%)	
	15%	8.1 (−21%)	7.4 (−7%)	5.7 (−44%)	5.7 (−41%)	6.6 (−35%)	
FCC	5%	8.2 (−20%)	8.6 (+8%)	8.3 (−5%)	6.8 (−30%)	-	[218]
	10%	7.3 (−28%)	7.5 (−6%)	7.1 (−18%)	5.8 (−40%)	-	
	15%	7.0 (−31%)	6.7 (−16%)	6.9 (−21%)	5.4 (−44%)	-	

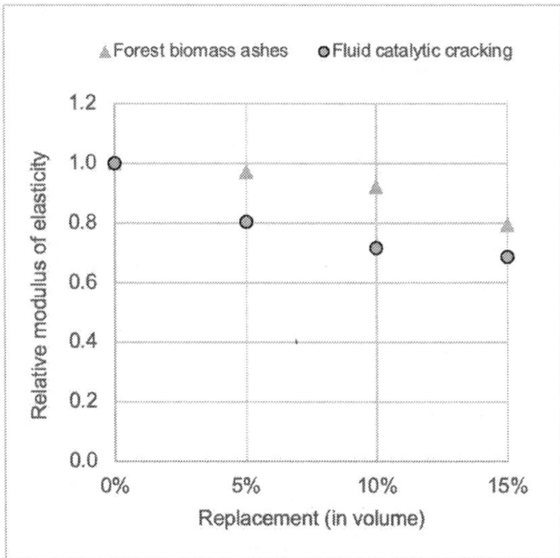

Fig. 5.40 Relative modulus of elasticity between mortars with binder wastes and their plain mortar (28 days) [215] [218].

natural aggregate). The binder wastes replaced cement at 5%, 10%, and 15%. The ratio of replacement for a mortar poorer in cement resulted in lower mechanical strength, as discussed in Sections 5.1.1 and 5.1.2. Similarly, the incorporation of these wastes decreased the modulus of elasticity. The decrease was as high as the volume of wastes incorporated. The incorporation of forest biomass ashes at 15% resulted in a decrease by 20% of the modulus of elasticity. The FCC waste, for the same volume of replacement, reduced this modulus by about 30%, both at 28 days. Martínez-Lage et al. [56] replaced CEM I cement with eucalyptus biomass ashes. The authors also found a decrease in the modulus of elasticity when the ashes replaced the binder. A 10% (wt.) replacement resulted in a decrease from 30 GPa to 29.8 GPa and 20% (wt.) replacement in a reduction to 27.5 GPa.

The reduction of modulus of elasticity indicates that these mortars are more deformable than the corresponding plain mortar. The binder wastes can have some pozzolanic properties but they do not have the same hydraulic properties as cement, and therefore the matrix is less compact than that of the reference mortars. This decrease in compactness is responsible for a lower modulus of elasticity and a higher deformability.

The reinforcement of mortars by the incorporation of fibers intends to improve the mortars' deformability and toughness. In Table 5.26, some studies' results about the influence of fiber wastes on the modulus of elasticity of render mortars are summarized. The relative modulus of elasticity between the reinforced mortars and their correspondent reference mortar is illustrated in Fig. 5.41.

The examples indicate that, in fact, the incorporation of fibers in render mortars improves their deformability, that is these mortars in general present lower modulus

Table 5.26 Modulus of elasticity of blended mortars with fiber wastes' incorporation obtained by other authors.

Type of fiber	Addition percentage	Characteristics	Modulus of elasticity (GPa)				References
			28 days	90 days	180 days	365 days	
Sheep's wool	10%	L: 1.5 cm	16.0 (−1%)	15.2 (−4%)	14.8 (−5%)	-	[126]
		L: 3.0 cm	12.8 (−21%)	11.7 (−26%)	11.4 (−27%)	-	
	20%	L: 1.5 cm	15.3 (−2%)	14.3 (−10%)	13.6 (−13%)	-	
		L: 3.0 cm	16.9 (+4%)	17.5 (+10%)	16.8 (+8%)	-	
Textile	0.125% (wt.)	-	9.0 (+12%)	-	-	-	[178]
	0.25% (wt.)	-	7.3 (−8%)	-	-	-	
	0.50% (wt.)	-	7.8 (−3%)	-	-	-	
	5%	L: 1.5 cm	8.6 (−16%)	7.6 (−5%)	7.0 (−20%)	6.6 (−32%)	[218]
		L: 3.0 cm	8.7 (−15%)	7.8 (−3%)	7.1 (−18%)	6.8 (−30%)	
	10%	L: 1.5 cm	12.7 (−22%)	12.3 (−23%)	-	-	[223]
		L: 3.0 cm	13.1 (−19%)	12.8 (−19%)	-	-	
	20%	L: 1.5 cm	10.9 (−33%)	11.0 (−31%)	-	-	
		L: 3.0 cm	11.6 (−28%)	11.5 (−28%)	-	-	
Acrylic	5%	L: 1.5 cm	8.4 (−18%)	8.3 (+4%)	7.5 (−14%)	7.8 (−20%)	[175]
		L: 3.0 cm	8.4 (−18%)	7.9 (−1%)	7.7 (−11%)	7.0 (−28%)	
Glass	5%	L: 1.5 cm	9.4 (−8%)	8.8 (+10%)	8.6 (−1%)	8.6 (−11%)	[175]
		L: 3.0 cm	9.4 (−8%)	8.3 (+4%)	8.3 (−5%)	7.3 (−25%)	

of elasticity than that of the reference mortar. Only the incorporation of sheep's wool fibers [126] and that of fibers of 0.125% (in weight) [178] resulted in increments of the modulus of elasticity. This reduction of modulus of elasticity, obtained by the frequency of resonance, can be due to the higher volume of water needs. Fibers, as discussed in Section 4, are more absorbent than natural aggregates or the binder, thus, when added to mortars they increase their water needs. This increment influences the

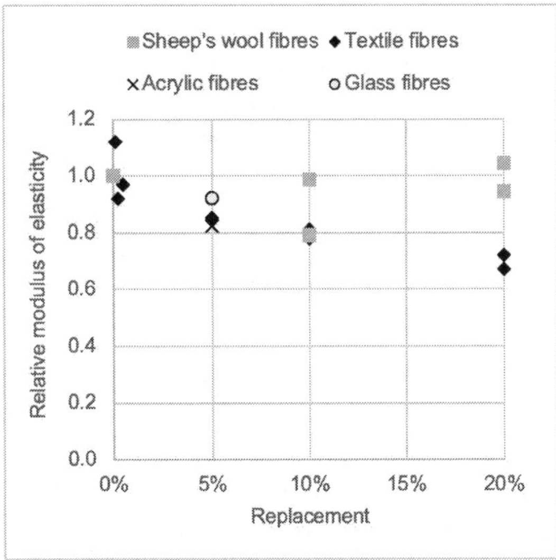

Fig. 5.41 Relative modulus of elasticity between mortars with fiber wastes and the corresponding plain mortar (28 days) [126] [218] [223].

mortar's porosity and consequently the mortar's modulus of elasticity. The improvement of deformability allows the render mortars to accommodate the induced stresses maintaining their integrity and avoiding cracking. The incorporation of fibers also significantly improves the toughness of the mortars, as presented in the following section.

The modulus of elasticity can be related to the mortars' compressive strength (as illustrated in Fig. 5.42). A mortar with a high modulus of elasticity normally has a high compressive strength. This relation is due to the mortar's compactness. A compact mortar presents high strength and simultaneously is less deformable. On the contrary, a weaker mortar is more deformable. Concerning the wastes and how they can influence the modulus of elasticity, from Fig. 5.42 it is noticed that each waste presents a different behavior, although all maintain a similar trend. For example, aggregate wastes, in fact, present lower compressive strength and higher deformability (lower modulus of elasticity); powder wastes, on the contrary, are the wastes that confer more strength to the mortars and provide lower deformability.

5.2.2 *Prefailure and postfailure behavior*

In this chapter, the behavior of the mortars before and after their maximum strength is achieved is discussed: prefailure and postfailure behavior.

In Fig. 5.43, typical curves of the load/displacement of render mortars (without waste incorporation), when a compressive load is applied, are presented. In the figure, three render mortars with distinct binder: aggregate ratios (1:4, 1:5, and 1:6)

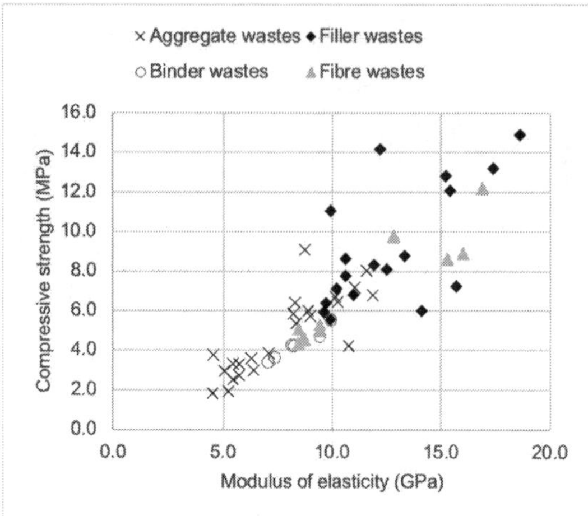

Fig. 5.42 Modulus of elasticity versus compressive strength [7] [8] [18] [49] [51] [126] [137] [142] [143] [148] [154] [155] [172] [180] [182] [215] [218] [223].

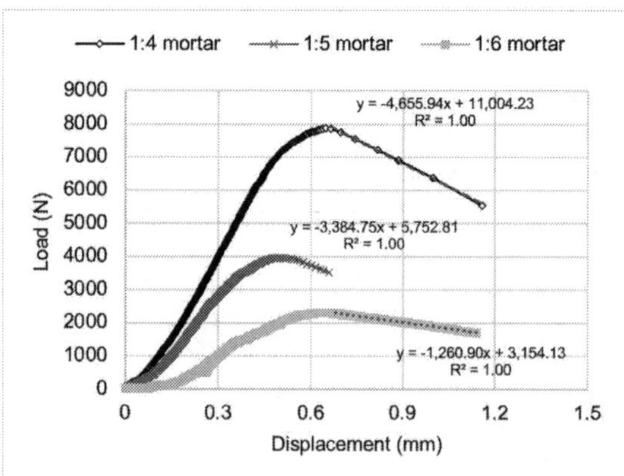

Fig. 5.43 Compressive strength development of mortars with volumetric ratios of 1:4, 1:5, and 1:6 (example) [1] [7].

are represented. From these curves, it is noticed that the mortar with higher volume of cement (1:4 mortar) is the one that withstands higher load. In this specific case, the mortar can withstand almost 8 kN, while the mortar with a volumetric ratio of 1:5 withstands 4 kN and the mortar poorer in cement (1:6) slightly more than 2 kN.

With the incidence of the load, the mortar 1:6 presents a major displacement, which indicates more prefailure deformability than the remaining ones. To evaluate the postfailure behavior of the mortar, the tangents to the force-displacement curves in the branch after the maximum force were drawn and their slope was calculated. As seen in Fig. 5.43, the slope of the mortar richer in cement (1:4) was -4655 N/mm and the slope of the tangent to the curve of 1:6 mortar was -1260 N/mm, considerably lower. Comparing these two slopes, it is noticed that the 1:4 mortar, after achieving its maximum strength, has a higher decay of strength than the mortar 1:6. In render applications that means the 1:6 mortar is still able to deform after failure, probably due to creating cracks with lower width and depth or not promoting the cracking opening in comparison with the 1:4 mortar.

The proportion between the binder and aggregates influences the compressive strength of the mortars, as well as other modifications to the mortar's porosity. Wastes' incorporation in render mortars is responsible for a modification in their microstructure and consequently affects pre- and postfailure behaviors.

From Figs. 5.44–5.47, several curves from render mortars with aggregate, filler, binder and fiber wastes with volumetric ratios of 1:4 (binders : aggregates), are represented. The figures allow concluding that filler wastes are the ones that lead to stronger mortars, while binder wastes or fiber wastes give rise to mortars with lower compressive strengths, in comparison with conventional mortars with no waste's incorporation. Aggregate wastes in some cases, for example for sanitary ware or CDW (mainly composed of concrete) incorporation, lead to stronger mortars than the reference mortar. In other cases, such as polymers (shredded tires, PF or PP), when these wastes are incorporated as aggregates in mortars, they have lower compressive strengths.

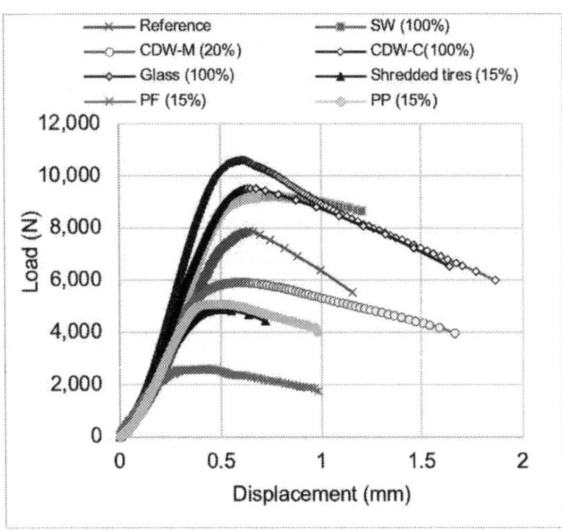

Fig. 5.44 Compressive strength development curves of mortars with aggregate wastes (examples). Extra data from other studies [137] [143] [148] [155].

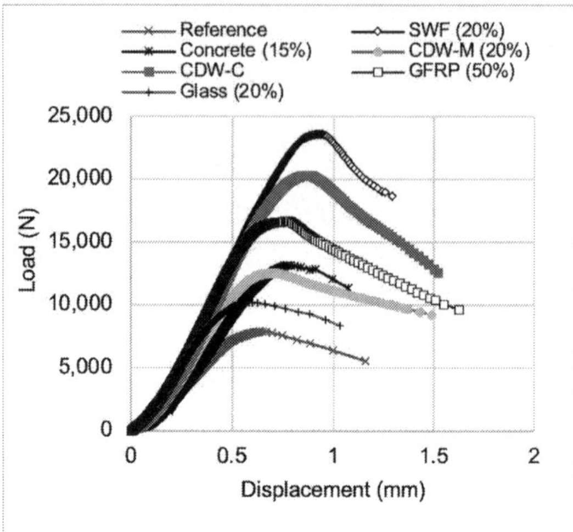

Fig. 5.45 Compressive strength development curves of mortars with filler wastes (examples) Extra data from other studies [7] [51] [142] [154] [172] [218].

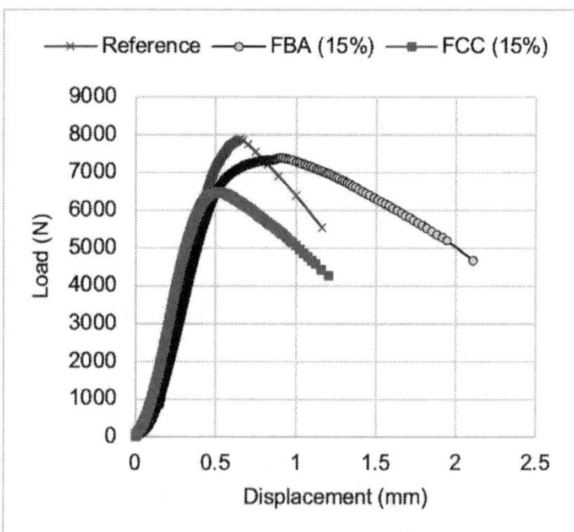

Fig. 5.46 Compressive strength development curves of mortars with binder wastes (examples). Extra data from another study [218].

From the compressive strength test and its load/displacement curves, it is possible, in addition to the maximum value of strength, to obtain information about the pre- and postfailure mortars' behavior. The development of the compressive strength curves, as the moment of the mortars' loading until failure is achieved, depends on the hardness and deformability of the specimens.

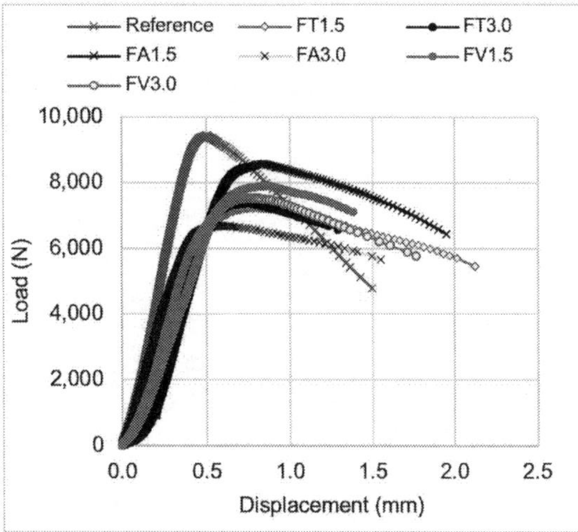

Fig. 5.47 Compressive strength development curves of mortars with fiber wastes (examples). Extra data from other studies [218] [223].

Concerning the prefailure behavior, in Fig. 5.44 it is noticed that some aggregate wastes' mortars present a higher modulus of elasticity than the reference mortar, namely glass or CDW-C waste mortars. Mortars with polymer wastes such as PF, PP, or shredded tires, on the other hand, have lower compressive strengths but are more deformable in prefailure than the conventional mortar. Filler wastes, as already noticed in the modulus of elasticity test, present a higher value than the reference mortar, which can be noticed in the Fig. 5.45. Mortars with binder wastes (Fig. 5.46), until failure, present a similar behavior to that of the reference mortar. Fiber wastes (Fig. 5.47) have a more deformable prefailure pattern, presenting a lower modulus of elasticity, as already discussed.

From the load/displacement curves, it is also possible to observe the postfailure behavior of the mortars (Figs. 5.44– 5.47). Brittle mortars present an abrupt decay of strength after the maximum load is achieved (postfailure), while more ductile mortars present a slighter decay. The tangent to the curves (after the maximum load was achieved) presented in Figs. 5.44– 5.47 was analyzed and their slope is summarized in Table 5.27. These slopes are lower for the aggregate wastes than for a reference mortar, with no wastes' incorporation. This means that these mortars in general present a better postfailure pattern than conventional mortar, being able to deform more after failure. The incorporation of plastics, for example PF and PP, is responsible for a slope reduction by more than 50%. On the other hand, mortars with filler wastes present more brittle behavior after failure than the reference mortar. Sanitary ware incorporation at 20%, for example, increases the slope fourfold, where the modified mortar is much more brittle than the reference mortar. The replacement of cement with wastes, for example forest biomass ashes or FCC waste, results in a more ductile postfailure behavior, with a lower tangent slope to the load/displacement curve, in

Table 5.27 Compressive strength behavior after failure is achieved—tangent to the curves slope.

Wastes type	Mortar	Replacement ratio	Slope after failure (N/mm)	Slope in comparison with reference mortar	R2	References
	Reference mortar	**0%**	**-4655.9**	**-**	**1.00**	**Extra data from [7]**
Aggregate	Sanitary ware	100%	−1308.5	Softer	0.97	Extra data from [18]
	CDW (predominantly concrete)	20%	−1837.5	Softer	0.99	Extra data from [155]
	CDW	100%	−3826.0	Softer	0.99	Extra data from [155]
	Glass	100%	−3044.8	Softer	0.99	Extra data from [143]
	Shredded tires	15%	−2051.2	Softer	0.96	Extra data from [148]
	PET flakes	15%	−1420.4	Softer	0.99	Extra data from [137]
	PET pellets	15%	−2183.8	Softer	0.99	Extra data from [137]
Filler	Sanitary ware powder	20%	−16005.3	Higher	0.98	Extra data from [7]
	Concrete powder	15%	−6564.0	Higher	0.95	Extra data from [51]
	CDW (predominantly concrete)	20%	−4219.2	Higher	1.00	Extra data from [154]
	CDW	20%	−12285.3	Higher	1.00	Extra data from [154]
	Glass powder	20%	−4340.6	Higher	0.98	Extra data from [142]
	GFRP powder	50%	−8412.6	Higher	0.99	[172] [218]
Binder	Forest biomass ashes	15%	−2277.4	Softer	0.99	[215] [218]
	FCC	15%	−3249.3	Softer	0.99	[218]
Fiber	TF1.5	5%	−1314.6	Softer	0.99	[218]
	TF3.0	5%	−1499.3	Softer	1.00	[218]
	AF1.5	5%	−1879.4	Softer	0.99	[218]
	AF3.0	5%	−1137.6	Softer	0.99	[218]
	GF1.5	5%	−1470.7	Softer	0.99	[218]
	GF3.0	5%	−1840.8	Softer	0.99	[218]

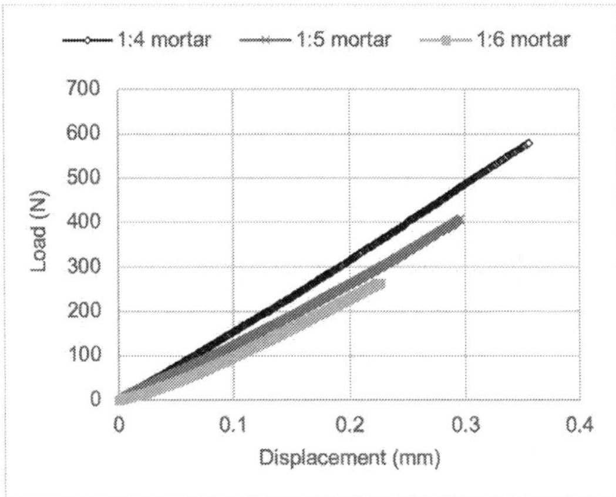

Fig. 5.48 Flexural strength development of mortars with volumetric ratios of 1:4, 1:5, and 1:6 (example) [1] [7].

comparison with the one of the reference mortar. An improvement of the pattern after the maximum load can also be noticed for fiber wastes.

In Fig. 5.48, load/displacement curves obtained in a flexural strength test of three reference render mortars are represented. The mortars present the same cement type and natural aggregate, only differing in the proportion between the binder and aggregates: 1:4, 1:5, or 1:6. The flexural strength decreases whenever the ratio between the binder and aggregates decreases. Thus, the 1:6 mortar is the one with lower flexural strength, but it is also the most deformable mortar. This deformability can be noticed by the development of the prefailure curve until failure. The mortar can deform more with the load incidence than the 1:4 mortar, which corresponds to a lower modulus of elasticity.

Similar to the modification of the binder: aggregate ratio, the incorporation of wastes also modifies the development of the load/displacement curves obtained during the flexural strength test. From Figs. 5.49–5.52, flexural strength curves of a reference mortar and mortars with aggregate, filler, binder, and fiber wastes are presented. All the mortars represented have a volumetric ratio of 1:4 (binders:aggregates). The load/displacement curves depend on the waste type incorporated.

Natural aggregates replacement by aggregate wastes can improve the flexural strength of the mortars. The angularity, roughness, and porosity of the waste particles can improve the interface between the wastes and the cementitious matrix and increment the flexural strength of the mortars. Glass, red ceramics, or concrete (from construction and demolition wastes) are examples of waste that can replace the natural aggregate and increment the mortars' flexural strength. Some wastes can simultaneously improve the flexural strength and the deformability of the mortars, as can be noticed in Fig. 5.49.

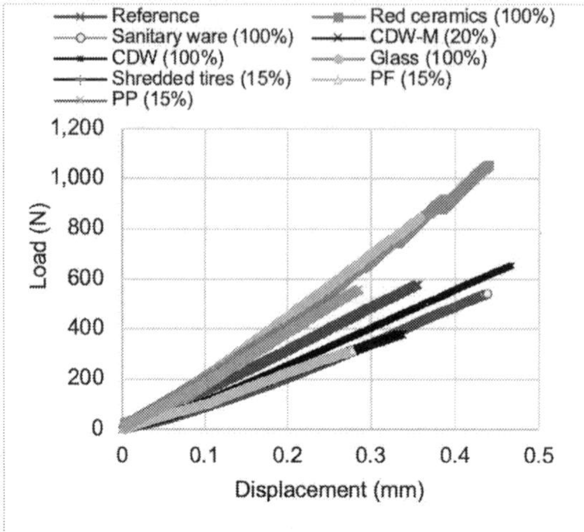

Fig. 5.49 Tensile strength development curves of mortars with aggregate wastes (examples). Extra data from other studies [7] [18] [137] [143] [148] [155] [221].

Fillers increase the compacity of mortars, as already noticed in several mortars' properties. This increment of compactness is responsible for higher strengths as was noticed in compressive strength and flexural strength. These increments of strength are also accompanied by higher modulus of elasticity and consequently lower deformability. In Fig. 5.50, it can be seen that the reference mortar presents a lower slope with the load incrementing, while the mortars with filler wastes present a steeper slope.

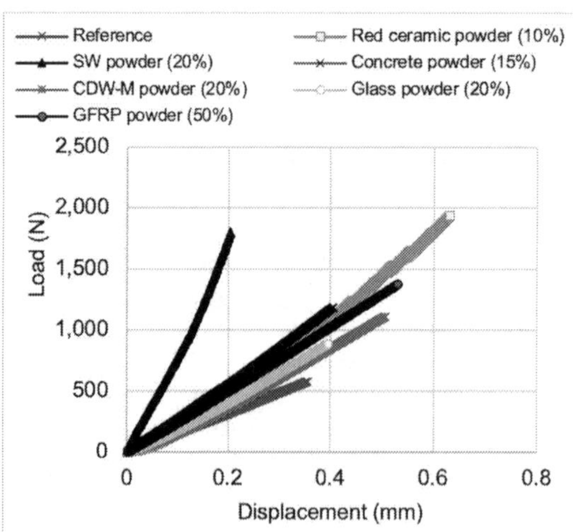

Fig. 5.50 Tensile strength development curves of mortars with filler wastes (examples). Extra data from other studies [7] [49] [51] [142] [154] [172] [218].

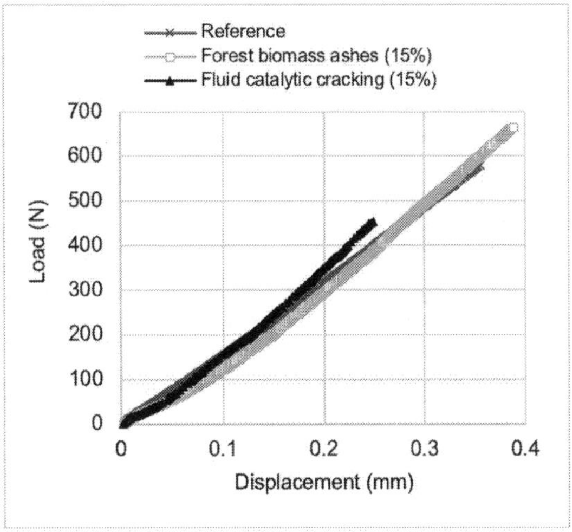

Fig. 5.51 Tensile strength development curves of mortars with binder wastes (examples). Extra data from other studies. [7] [218] [223].

Sanitary ware waste stands out from the other wastes; although it is one of the most resistant wastes, it is also the one that achieves less displacement with the load incidence, meaning that it is the waste that provides lower deformability to the mortars.

Two examples of binder wastes are presented in Fig. 5.51. Generally, binder wastes decrease the mortars' strength. These wastes can have some pozzolanicity. However, they do not have the same properties as cement, and the reduction of strength is a natural effect of the use of binder wastes. Nevertheless, the incorporation of these wastes improves the deformability of the mortars, as presented in Section 5.2.1. Elastic deformability.

Fibers usually have more influence on flexural strength than on compressive strength, as they act as bridges between the different points of the matrix. When a flexural load is applied, the fibers are meant to function as staples when the first crack occurs, providing higher flexural strength and delaying failure. On the other hand, the fibers properties, namely modulus of elasticity and tensile strength, must be carefully chosen to be favorable to the improvement of the mortars' flexural strength. The fibers must present a modulus of elasticity moderately higher than the mortar's, for them to be mobilized. If this modulus is smaller, the fibers are more deformable than the matrix and they deform without creating restriction to the matrix leading to the mortars' failure. In that case, the fibers are not effective and do not promote the bridging effect. Thus, it is necessary for the fiber to present a slightly higher modulus of elasticity than the mortars' but of the same order of magnitude. On the other hand, it is essential that the tensile strength of the fibers is higher than the mortar's to reinforce them. If the tensile strength is lower than the mortar's and the modulus of elasticity is higher or of the same order of magnitude, the fiber will achieve failure before the mortar. In Fig. 5.52, load/deflection curves of four mortars reinforced with textile fibers (illustrated in Fig. 5.53) are represented. Two lengths and two addition

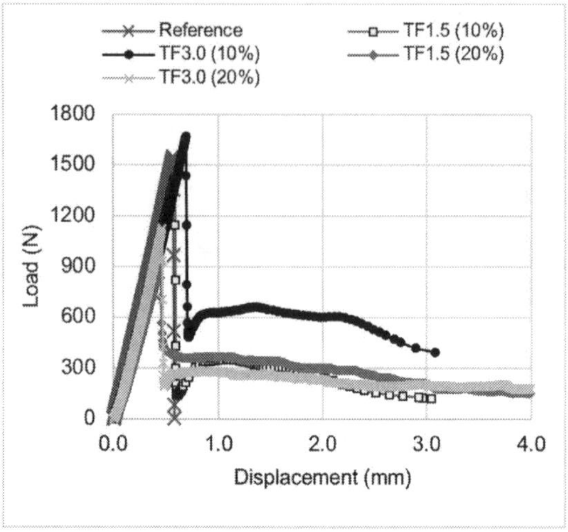

Fig. 5.52 **Tensile strength development curves of mortars with fiber wastes (examples).** Extra data from other studies [7] [218] [223].

Fig. 5.53 Illustration of textile fibers waste.

percentages are represented. Fibers can reduce the flexural strength of render mortars. This decrease is related to the water absorption of the wastes, which was discussed in the chapter relating to the fresh properties of mortars. Nevertheless, the incorporation of fibers has a great influence on the mortars' behavior whenever loaded. In the figure, it is visible that the mortar, after reaching the peak load, is still able to withstand a smaller load and deflect for a period of time, presenting a better postfailure behavior than that of the reference mortar. In render mortars, this corresponds to a reduction of the cracks widths whenever the maximum load is achieved and to a reduction of the cracking propagation.

5.3 Chapter conclusions

Render mortars may be resistant to several external actions without cracking. These actions include thermal variations, humidity variations, ice formation, blows, impacts, movements of the substrate, or others. Mortars, to resist to these actions, may present some strength and deformability. These three properties are essential and a balance between them is necessary to achieve adequate render mortars.

In this chapter, the influence of each type of waste in mortars' mechanical strengths (compressive, tensile, adherence, and impact strengths) was discussed. Some examples of research were also included to sustain the conclusions reached. Briefly it was indicated that filler wastes significantly improve the mechanical strengths of mortars, for all the strength types considered. Aggregate wastes, in some cases, are also responsible for increments in strength, namely due to their shape and roughness. Binder wastes, on the contrary, reduce the mechanical strength of the mortars, in general. Fiber wastes in some cases can also improve the flexural strength.

Render mortars should be mechanically resistant, but if they are not deformable, they will crack when submitted to several types of stresses. In this sense, the influence of wastes on mortars' deformability was also discussed in this chapter. As expected, wastes that present higher strengths are also the ones that have lower deformability and a worse postfailure behavior. In this sense, filler wastes are responsible for mortars becoming more brittle while the remaining wastes are responsible for more ductile mortars. This ductility promotes a more favorable cracking behavior. The incorporation of fiber wastes is responsible for improvements in terms of postfailure pattern, being able to support some load after the maximum strength was achieved. These improvements are, however, dependent on the compatibility between the mortars' matrix and each type of fiber. The fibers can prevent cracking and, whenever cracks appear, these fibrous materials are able to minimize their widths and propagation.

Mortars with incorporation of wastes: water behavior

6

Renders and plaster are mortars applied on vertical walls that are interior or exterior, respectively. In both cases, these mortars are sacrificial layers applied aiming at protecting the substrate from external actions, increasing their lifetime. Ingress of liquid water is one of these external actions that can affect the durability of the substrate.

Liquid water can come from different sources, rain and soil moisture being the most common ones. Usually, a finishing layer is applied on top of the renders, which has specific characteristics, namely it can prevent the liquid water from reaching the render mortar. However, due to the inexistence of the finishing layer, or to its intrinsic characteristics or due to several anomalies, such as cracking of the finishing layer, the render mortar may be exposed to liquid water. In such cases, the render should prevent the water from migrating to the substrate.

A mortar, after absorbing water, should be able to release that amount of water, as quickly as the environmental conditions allow it. This water release occurs in the vapor state. In this chapter, the different mechanisms of water absorption and water release are discussed. Some results from several studies for mortars with incorporation of wastes are also presented, as well as the influences of the wastes in the mortars' water behavior.

6.1 Water penetration

Render mortars can absorb liquid water by capillarity or under pressure, depending on whether there is pressure on the surface of the mortars. Renders that are in contact with soil, and have not a barrier to water absorption, absorb water by capillarity in height (Fig. 6.1). The height that the water achieves depends on the porosity and diameter of the mortar pores. When the source of water is rain, it can be absorbed by capillarity or under pressure, depending on the pressure that the wind exerts on the external wall or on other sources of pressure, such as horizontal architectonic elements that allow accumulation of a given height of water. Next, the two types of water absorption and the influence of wastes in this absorption are presented.

6.1.1 Water absorption by capillarity

Water absorption can be evaluated by the curves of water absorption over time. A render mortar can be in contact with water for a short period of time, as in a daily raining day, or can be in contact for a longer period. These long periods can be consecutive raining days or water from soil sources.

Eco-efficient Rendering Mortars. DOI: 10.1016/C2018-0-04147-X

Fig. 6.1 Water absorption by capillarity rising from the soil (examples).

Mortars usually absorb a high volume of water in the first hours of exposure and then the rate of water absorption slows down. The curves of water absorption are, in this sense, composed of two approximately straight-line segments, as exemplified in Figure 2.8 (Chapter 2). The first one represents the initial period of water absorption and the other the absorption over time. The slope of the first straight-line segment is designated as "capillary coefficient." This coefficient is a mathematic tool to measure the speed of water absorption in the first minutes of water exposure. This coefficient is influenced by the diameter of the pores of the mortars. Renders with larger pores absorb in the same period a higher volume of water than smaller pores. From the curves, it is also possible to quantify the total volume of water absorption. This volume is given by the total amount of water absorbed during the test.

The total amount of water absorbed by capillarity depends on the total volume of capillary pores. The capillarity coefficient measures the water absorption rate in the first period of the test; therefore, for this coefficient, the larger pores (within the capillary range) are the ones with greater influence. The total water absorption by capillarity, on the contrary, involves all the capillary pores.

In Fig. 6.2, the typical absorption curves of reference mortars (without wastes and without admixtures) with different volumetric ratios are presented. The curves are identical, being composed of two slopes. The first straight line, which corresponds to the initial volume of water absorbed, lasts until 60 min in the case of 1:6 mortar and until 5 h in the case of 1:4 mortar. The time of water exposure, when the transition between the first and the second phase of absorption (represented by the straight lines) occurs, depends on the mortars and on their porosity. A mortar with a binder: aggregates volumetric proportion of 1:4 is more compact than a mortar with a ratio of 1:6. Consequently the mortar with the lower ratio (1:6) absorbs water faster than the mortar with the higher ratio (1:4).

The incorporation of wastes in mortars modifies their porosity and consequently their water absorption. In the following figures (Figs. 6.4–6.7), the development of the water absorption over time of mortars with incorporation of wastes, based on the different literature studies, is represented. In the studies presented next, the absorption by capillarity was evaluated in mortars' semiprisms that were cut in half for the

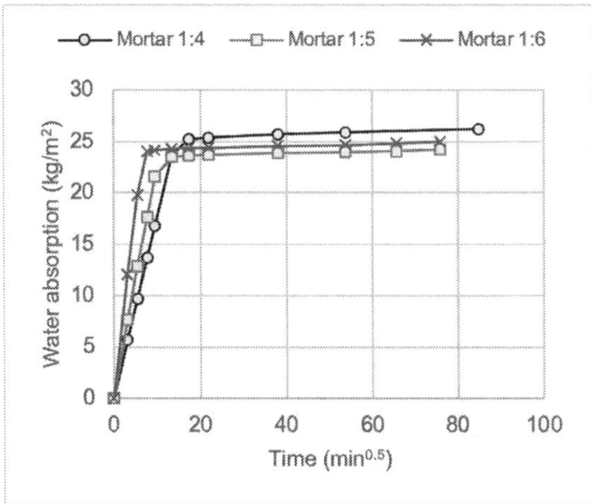

Fig. 6.2 Example of water absorption curves of mortars with volumetric ratios of 1:4, 1:5, and 1:6 [1].

Fig. 6.3 Water absorption by capillarity evaluated in laboratory (example).

test (80 × 40 × 40 mm). The specimens were sealed on the side faces to guarantee the capillary absorption and the sawn face was put in contact with water, as illustrated in Fig. 6.3. The development of these curves depends on the type of waste incorporated.

The water absorption curves of mortars with incorporation of aggregate wastes, as illustrated in Fig. 6.4, present a development above the curve of a reference mortar, meaning that these types of mortars absorb a higher volume of water than a mortar without wastes. The increment of water absorption is related to the volume of capillaries of these modified mortars. Previously, it was noticed that the incorporation of aggregate wastes, in general, was responsible for a water/binder ratio increment, due

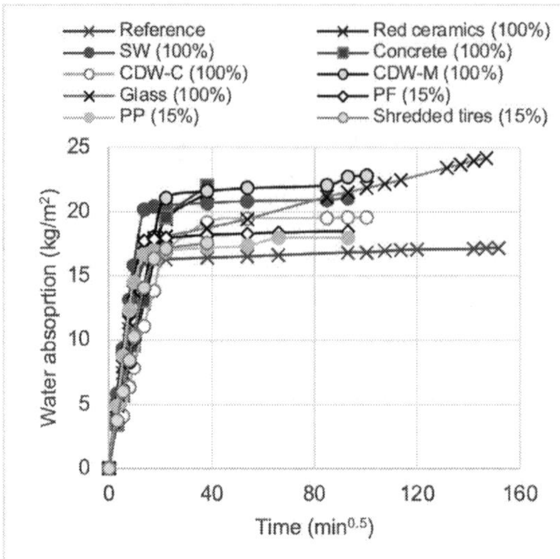

Fig. 6.4 Example of water absorption curves of mortars with aggregate wastes [8] [18] [137] [143] [148] [155] [221].

to their properties (as shape, roughness, porosity, and water absorption). This ratio increments the volume of mortar voids and consequently their capillaries. A mortar in contact with liquid water can absorb it by capillarity until the volume of capillaries is filled. The mortars with aggregate wastes present a higher volume of capillaries, absorbing greater water volume (Table 6.2). Besides absorbing a higher volume of water, the waste aggregates' mortar absorbs it faster than a reference mortar (without wastes). This rate of absorption is quantified by the coefficient of capillary water absorption, which corresponds to the slope of the first straight line. The aggregate wastes' incorporation increments this coefficient. In Table 6.1, the capillary coefficients of rendering mortars with incorporation of this type of wastes are summed up. The capillary coefficient depends mainly on the diameter of the capillaries. A mortar characterized by large diameter capillaries will absorb a higher volume of water faster than a mortar with the same volume of pores but with smaller diameters. For example, considering the analyzed studies, the incorporation of sanitary ware waste increments the total volume of absorption by 24% (16.7 kg/m^2 vs 20.7 kg/m^2) and the capillary coefficient by 41% (1.12 kg/(m^2•min$^{0.5}$) vs 1.58 kg/(m^2•min$^{0.5}$)), in comparison with a reference mortar, with no wastes incorporation.

The incorporation of filler wastes produces a different pattern than the aggregates' waste incorporation. The curves of water absorption appear below the reference mortar curve (Fig. 6.5). The total water absorption, in general, is lowered when filler wastes are incorporated. The water absorption speed, analyzed by the first straight-line slope, is also reduced with this wastes' incorporation depending on the type. In Table 6.1, it is noticed that, for all the references analyzed [7] [49] [51] [142] [154] [172], the capillary coefficient was reduced by the incorporation of this thin waste

Table 6.1 Capillary coefficient of mortars with incorporation of wastes ($kg/(m^2 \cdot min^{0.5})$).

Type of incorporation	Waste	0%	5%	10%	15%	20%	40%	50%	100%	References
Aggregate	Red ceramic	0.82	-	-	-	0.68	-	0.74	0.93	[221]
	Sanitary ware	1.12	-	-	-	1.13	-	1.18	1.58	[18]
	Concrete	1.27	-	-	-	1.24	-	1.25	0.96	[8]
	Construction and demolition waste (predominantly concrete)	0.90	-	-	-	0.90	-	0.90	0.90	[155]
	Construction and demolition waste (mixture)	0.90	-	-	-	0.80	-	1.00	1.20	[155]
	Glass	1.07	-	-	-	1.13	-	1.23	1.40	[143]
	Shredded tires	1.07	0.97	1.14	1.03	-	-	-	-	[148]
	Plastics (PET (Polyethylene terephthalate) flakes)	1.34	1.36	1.55	1.48	-	-	-	-	[137]
	Plastics (PET pellets)	1.34	1.35	1.45	1.52	-	-	-	-	[137]
Filler	Red ceramics powder	0.82	0.57	0.48	-	-	-	-	-	[49]
	Sanitary ware powder	1.12	-	0.75	0.66	0.78	-	-	-	[7]
	Concrete powder	1.27	0.91	0.76	0.71	-	-	-	-	[51]
		0.80	0.73	0.77	-	0.81	0.81	-	-	[9]
	Construction and demolition waste powder (predominantly concrete)	0.87	-	0.56	0.52	0.47	-	-	-	[154]
	Construction and demolition waste powder (mixture)	0.87	-	0.51	0.57	0.72	-	-	-	[154]
	Glass powder	1.57	-	0.97	0.91	0.76	-	-	-	[142]
	GFRP (Glass fibre reinforced polymer) powder	1.45	-	0.27	0.22	0.25	-	0.20		[172]
Binder	Forest biomass ashes	1.45	1.14	1.47	1.50	-	-	-	-	[215]
	Fluid catalytic cracking	1.45	1.29	1.34	1.44	-	-	-	-	[218]
Fibre	Textile fibres (1.5 cm long)	1.45	1.70	-	-	-	-	-	-	[218]
	Textile fibres (3.0 cm long)	1.45	1.65	-	-	-	-	-	-	[218]
	Acrylic fibres (1.5 cm long)	1.45	1.51	-	-	-	-	-	-	[218]
	Acrylic fibres (3.0 cm long)	1.45	1.61	-	-	-	-	-	-	[218]
	Glass fibres (1.5 cm long)	1.45	1.43	-	-	-	-	-	-	[218]
	Glass fibres (3.0 cm long)	1.45	1.52	-	-	-	-	-	-	[218]

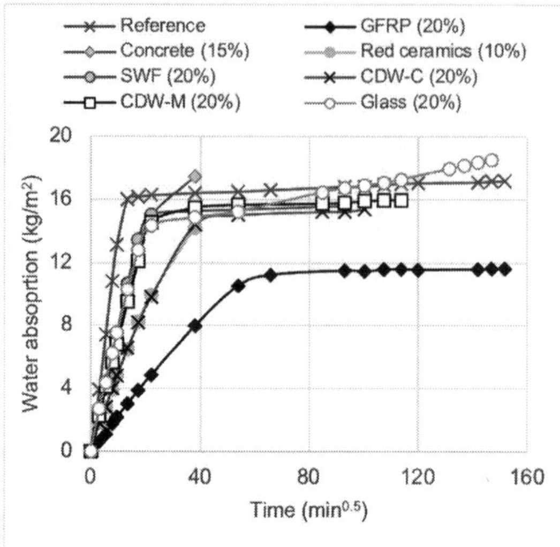

Fig. 6.5 Example of water absorption curves of mortars with filler wastes (fillers) [7] [49] [51] [142] [154] [172].

particles. The lowest value obtained occurred for the GFRP (Glass Fibre Reinforced Polymer) powder waste that reduced this coefficient by more than 80% compared to a reference mortar. However, over time, the total amount of water absorbed is higher than the one absorbed by the reference mortar [51] [142]. This phenomenon can be explained by the distribution of pores and their diameter. Filler materials intend to reduce the volume of the largest mortar pores, decreasing the average pore diameter. However, if the particles are in excess, the filler potential is exhausted, and the particles start to create empty voids among themselves. These particles reduce the diameter of the pores, but they can increase the sample voids' total volume. This results in slower absorption in the first period of water exposure but in higher volume of water absorption if the time of exposure is prolonged.

Two examples of water absorption curves of binder wastes (forest biomass ashes [215] and fluid catalytic cracking [218]) are represented in Fig. 6.6. The development of the curves is similar to that of the reference mortar. The incorporation of other byproducts that act as binders, in the mortars where cement was replaced by these wastes, seemed not to have a preponderant influence on the rate and volume of water absorption. The incorporation of forest biomass ashes (replacing cement by 15% in volume) resulted in more than 2% of water absorption total volume and more 3% of capillary coefficient than the reference mortar, which is not relevant.

In Fig. 6.7, the water absorption curves of three waste fibers (textile, acrylic, and glass) with two lengths (1.5 cm and 3.0 cm long) are represented. It is possible to observe that the development of the curves is similar between the mortars with fibers and the reference mortar. Although small differences can be noticed, it is possible to see that mortars with fibers, in general, absorb more water and faster

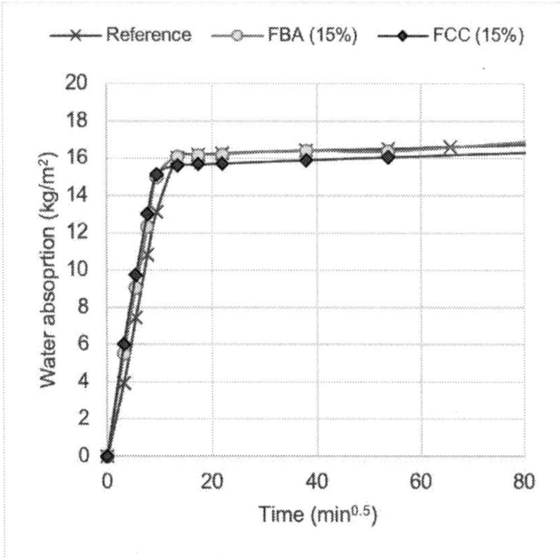

Fig. 6.6 Example of water absorption curves of mortars with binder wastes [215] [218].

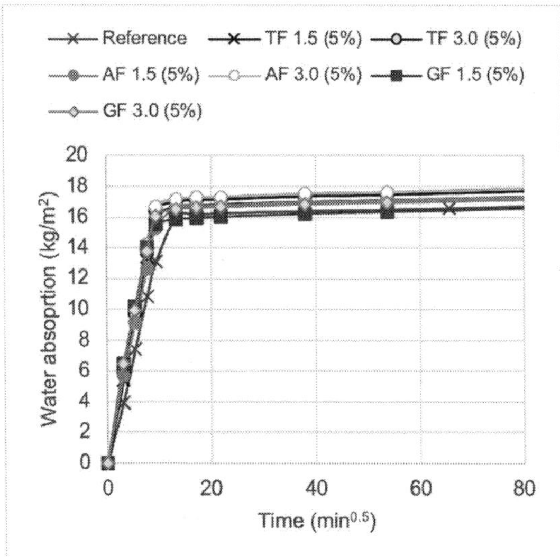

Fig. 6.7 Example of water absorption curves of mortars with fiber wastes [218].

than the reference mortar (Tables 6.1 and 6.2). For example, Farinha [218] noticed that textile fibers (1.5 cm long) absorbed 4% more water and 17% faster in the first period of absorption. It is important to mention that the values obtained in total water absorption depend on the volume of the specimen tested. In the studies presented in Table 6.1, the specimens used were semiprisms with 80 × 40 × 40 mm.

Table 6.2 Total water absorption of mortars with incorporation of wastes (kg/m^2).

Type of incorporation	Waste	0%	5%	10%	15%	20%	50%	100%	References
Aggregate	Red ceramic	17.0	-	-	-	16.9	18.2	21.4	[221]
	Sanitary ware	16.7	-	-	-	17.4	17.4	20.7	[18]
	Concrete	17.5	-	-	-	22.0	20.9	22.0	[8]
	Construction and demolition waste (predominantly concrete)	16.1	-	-	-	16.8	18.3	20.7	[155]
	Construction and demolition waste (mixture)	16.1	-	-	-	16.3	18.0	22.8	[155]
	Glass	19.8	-	-	-	20.5	21.3	24.2	[143]
	Shredded tires	17.7	17.2	17.4	17.6	-	-	-	[148]
	Plastics (PET flakes)	16.0	16.8	18.6	18.5	-	-	-	[137]
	Plastics (PET pellets)	16.0	17.1	17.5	18.0	-	-	-	[137]
Filler	Red ceramics powder	17.0	13.9	14.2		-	-	-	[49]
	Sanitary ware powder	16.7	-	15.5	15.3	15.6	-	-	[7]
	Concrete powder	17.5	17.6	16.9	17.4	-	-	-	[51]
	Construction and demolition waste powder (predominantly concrete)	16.1	-	14.0	15.2	16.0	-	-	[154]
	Construction and demolition waste powder (mixture)	16.1	-	16.0	15.0	15.3	-	-	[154]
	Glass powder	20.0	-	18.6	18.7	18.5	-	-	[142]
	GFRP (Glass Fibre Reinforced Polymer) powder	17.3	-	10.8	10.9	11.7	16.6	-	[172]
Binder	Forest biomass ashes	17.3	15.5	17.7	17.7	-	-	-	[215]
	Fluid catalytic cracking	17.3	16.4	16.5	17.0	-	-	-	[218]
Fibre	Textile fibres (1.5 cm long)	17.3	18.0	-	-	-	-	-	[218]
	Textile fibres (3.0 cm long)	17.3	18.4	-	-	-	-	-	[218]
	Acrylic fibres (1.5 cm long)	17.3	18.0	-	-	-	-	-	[218]
	Acrylic fibres (3.0 cm long)	17.3	18.6	-	-	-	-	-	[218]
	Glass fibres (1.5 cm long)	17.3	17.2	-	-	-	-	-	[218]
	Glass fibres (3.0 cm long)	17.3	17.8	-	-	-	-	-	[218]

In fact, water absorption curves depend on the type of waste incorporated, both in terms of the total volume of water absorbed and the rate of that absorption. In Figs. 6.8 and 6.9, the results obtained by some studies about the incorporation of aggregate, filler, binder,

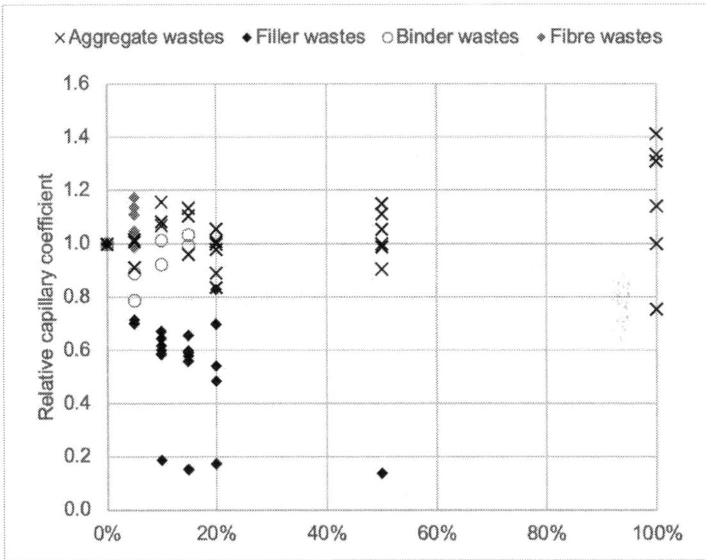

Fig. 6.8 Relative capillary coefficient of mortars with aggregate, filler, binder, or fiber wastes [7]–[9] [18] [49] [51] [137] [142] [143] [148] [154] [155] [172] [215] [219] [221].

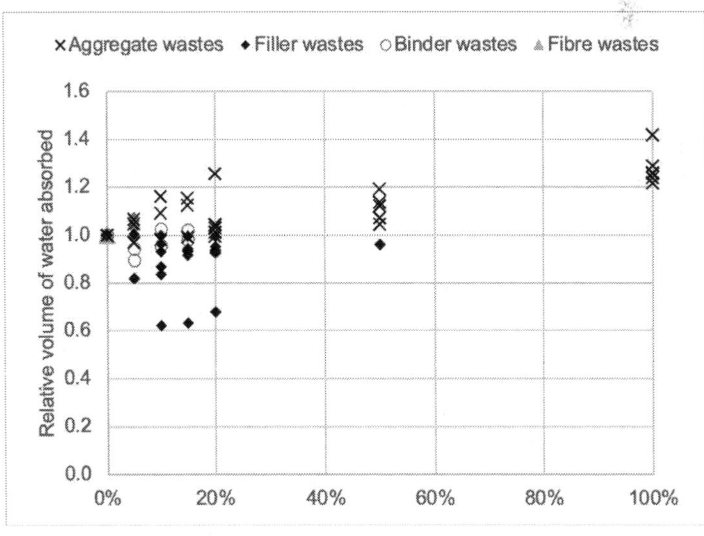

Fig. 6.9 Relative total volume of water absorption of mortars with aggregate, filler, binder, or fiber wastes. [7] [8] [18] [49] [51] [137] [143] [148] [154] [155] [172] [215] [218] [221].

and fiber wastes in the capillary coefficient and in the total amount of water absorption, in comparison with reference renders with no incorporation of wastes, are summarized.

From Figs. 6.8 and 6.9, it can be clearly noticed that filler wastes reduce the amount and speed of water absorption. The capillary coefficient is reduced with the volume of filler wastes incorporated. The total amount of water absorption, although it is in general lower than the correspondent reference mortars, increases with the volume of wastes incorporated. This increment is associated to what was mentioned as the exhaust of the filler effect. The filler wastes fill the voids between the particles and start to create voids among themselves, increasing the volume of pores of the material. Aggregates, on the contrary, are the type of incorporation that absorbs more water and faster than the reference mortar. From Figs. 6.8 and 6.9, it can be noticed that the water absorption increases with the volume of wastes incorporated. In those figures, it is also noticeable that the incorporation of binder wastes or fiber wastes has lesser influence in water absorption than that of aggregate or filler wastes.

6.1.2 Permeability to water under pressure

Water may be in contact with render mortars with some pressure. Wind is an example of pressure that is applied to external walls. Renders should prevent the water from reaching deeper layers, in particular the substrate.

The volume of water absorbed by the mortar not only depends on their capillary pores range, as the water absorption by capillarity, but also it involves larger pores and cracks. In Fig. 6.10, an example of evaluation of water absorption by a render mortar, under a given pressure, is illustrated.

Fig. 6.10 Water absorption under pressure test (example).

The incorporation of wastes as byproducts in render applications influences the porosity of the mortars and their pores distribution and, consequently, their behavior in water absorption under pressure. In the literature, some studies can be found about the influence of wastes in this type of water absorption. In Fig. 6.11, the water penetration under pressure of aggregate, filler, binder, and fiber wastes is presented. The values obtained presented in the figure are adimensional and correspond to relative values obtained in relation to the reference mortar of each study.

Aggregate wastes, as previously described, are known to increase the porosity of the mortars. Their angular shape, due to resizing treatments, their irregularity and roughness, allied to higher water absorption, are responsible for the porosity of the mortars. Due to this porosity, in Fig. 6.11, for aggregate wastes where sanitary ware and construction demolition wastes are included, an increment of water absorption under pressure is noticed. This increment of water absorption is in general higher as the volume of incorporation increases. Mortars with construction and demolition waste (a mixture of wastes such as ceramics, concrete, glass and others) absorb almost 80% more water than a reference mortar, when replacing sand at 20% [155]. In the case of sanitary ware, on the other side, even replacing sand at 100% only produces a water absorption 60% higher than that of the reference mortar (exclusively with sand as an aggregate) [218].

Filler wastes, on the other side, are responsible for reducing the porosity of the mortars due to a "filler effect." This reduction of porosity has as a consequence the reduction of water absorption under pressure (Fig. 6.11 and Table 6.3). This reduction depends on the waste and can vary between 20% and 80%. Construction and demolition wastes, in general, when incorporated as fillers at 20%, reduce the water

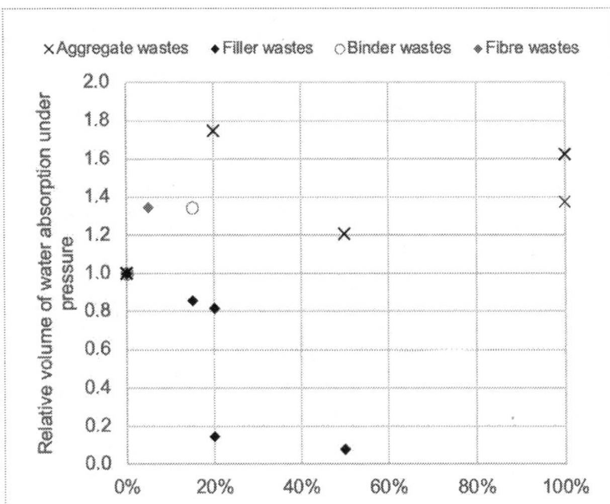

Fig. 6.11 Relative volume of water absorption under pressure by mortars with aggregate, filler, binder, or fiber wastes [7] [18] [154] [155] [172] [215] [218].

Table 6.3 Permeability to water under pressure of mortars with incorporation of wastes (mL/cm^2).

Type of incorporation	Waste	0%	5%	15%	20%	50%	100%	References
Aggregate	Sanitary ware	3.74	-	-	-	4.51	-	[18]
		9.20	-	-	-	-	12.63	[218]
	Construction and demolition waste (predominantly concrete)	2.17	-	-	-	-	3.52	[155]
	Construction and demolition waste (mixture)	2.17	-	-	3.79	-	-	[155]
Filler	Sanitary ware powder	3.74	-	-	0.54	-	-	[7]
	Construction and demolition waste powder (predominantly concrete)	2.17	-	-	1.77	-	-	[154]
	Construction and demolition waste powder (mixture)	2.17	-	1.86	-	-	-	[154]
	GFRP (Glass Fibre Reinforced Polymer) powder	9.20	-	-	-	0.72	-	[218]
Binder	Forest biomass ashes	9.20	-	12.36	-	-	-	[218]
Fibre	Textile fibres (1.5 cm long)	9.20	12.36	-	-	-	-	[218]

under pressure absorption by about 20% [155]. Sanitary ware, for the same volume of replacement, reduces the absorption under pressure by more than 80% [7]. Glass fiber–reinforced polymer powder, a polymeric waste, when replacing sand at 50%, reduces the absorption under pressure of the mortars by about 90% [218].

Binder wastes are byproducts supposed to replace cement by presenting some agglomeration ability. This type of wastes does not usually present the same properties of the usual binders, resulting in more porous mortars. Although a small influence can be noticed in capillary absorption, when the water is absorbed under pressure a major influence is found. This fact is due to the range of pores. The replacement of cement by binder wastes can result in mortars with large pores but with a similar volume of capillary pores. For example, forest biomass ashes replacing cement by 15% absorb almost 35% more mass of water under pressure [218].

Mortars with fiber wastes, similar to those with binder wastes, present a higher water absorption under pressure than the reference mortar [218]. In Fig. 6.11, water absorption under pressure of a mortar reinforced with 1.5-cm-long textile fibers is represented. The addition of the fiber waste results in an absorption under pressure almost 35% higher than the same mortar without additions. Fiber wastes are responsible for an increment in the porosity of the mortars, due to their threads, irregularities, and due to capability to retain water in their surface.

6.2 Water release

Water is one of the most important factors of degradation for both renders and masonry. When water is absorbed, the render should dry easily to prevent associated degradation problems such as ice formation in its pores, formation of efflorescence, biocolonization and, in extreme cases, when water reaches the masonry, also degradation of the substrate. Therefore, water absorption and drying tests are fundamental to characterize the render's performance.

6.2.1 Permeability to water vapor

The permeability to water vapor is manifested by the ability of the mortars to release the amount of water vapor contained in the mortars' pores. This permeability is, as well as the water absorption (by capillarity or under pressure), directly related to the porosity of the mortars. Porous mortars, with a high volume of interconnected pores, present a higher permeability to water vapor and release this amount of water faster than a more compact mortar. For the same total volume of pores, mortars with larger pores present higher vapor permeability than a mortar with smaller diameter pores.

The incorporation of wastes, instead of natural aggregates, binders or both, modifies the microscopy of the mortars, namely their porosity and consequently their permeability to water vapor.

Permeability to water vapor is often evaluated by the equivalent air thickness. This thickness corresponds to the air layer that presents the same resistance to water vapor diffusion than the material analyzed [226]. In Table 6.4, equivalent air thickness

Table 6.4 Equivalent air thickness of mortars with incorporation of wastes (m) tested with 2 cm thickness.

Type of incorporation	Waste	0%	10%	15%	20%	50%	100%	References
Aggregate	Red ceramics	0.16	-	-	-	0.10	-	[221]
	Sanitary ware	0.25	-	-	-	0.24	-	[18]
	Concrete	0.16	-	-	0.15	-	-	[8]
	Construction and demolition waste (predominantly concrete)	0.21	-	-	-	-	0.15	[155]
	Construction and demolition waste (mixture)	0.21	-	-	0.16	-	-	[155]
	Shredded tires	0.09	-	0.08	-	-	-	[148]
	Plastics (PET flakes)	0.17	-	0.15	-	-	-	[137]
	Plastics (PET pellets)	0.17	-	0.16	-	-	-	[137]
Filler	Red ceramic powder	0.16	0.18	-	-	-	-	[49]
	Sanitary ware powder	0.25	-	-	0.33	-	-	[7]
	Concrete powder	0.16	-	0.10	-	-	-	[51]
	Construction and demolition waste powder (predominantly concrete)	0.21	-	-	0.24	-	-	[154]
	Construction and demolition waste powder (mixture)	0.21	-	0.22	-	-	-	[154]
	GFRP (Glass Fibre Reinforced Polymer) powder	0.20	-	-		0.44	-	[218]

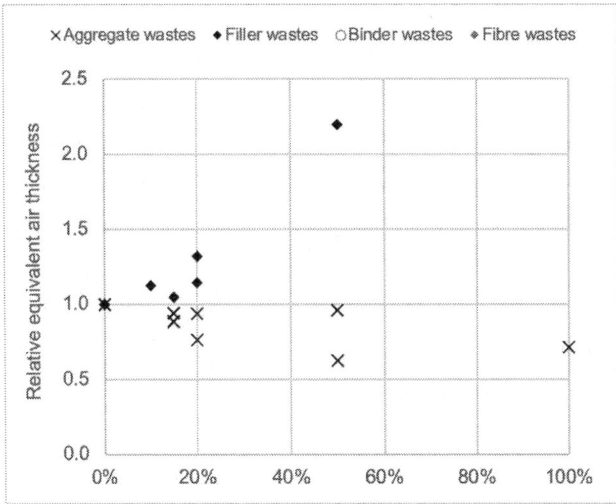

Fig. 6.12 Relative equivalent air thickness of mortars with aggregate, filler, binder, or fiber wastes [7] [8] [18] [49] [51] [137] [148] [154] [155] [218] [221].

values obtained by some authors for mortars with wastes incorporation are presented. In Fig. 6.12, the relative equivalent air thickness between modified mortars and reference mortar is also shown. In the figure, filler and aggregate wastes replacing sand are represented. The behavior of the wastes is opposite. Filler wastes increment the equivalent air thickness and aggregate wastes decrease it, both in comparison with the respective reference mortars. The higher the volume of filler wastes is the higher equivalent thickness. On the other side, the higher the volume of aggregate wastes is, the lower this thickness. This phenomenon is related to the porosity of the mortars. Mortars with aggregate wastes, more porous than the reference mortar, present a higher permeability to water vapor, which results in a lower equivalent air thickness. Fillers, on the other side, reducing the porosity of the mortars, produce mortars less permeable to water vapor and with a higher equivalent air thickness.

6.2.2 Unidirectional drying

Drying is the materials' ability to release the amount of water contained in their pores. A porous material dries more easily than a less porous one. The volume of pores and the pores' size influence the drying process, as larger pores loose water more easily than smaller ones, due to the stresses generated in the pores surface.

The porosity of render mortars depends on the type of binder, binder/aggregates ratio, and type of aggregate (more or less porous).

Drying behavior can be analyzed by the permeability to water vapor, as presented previously. However, drying also occurs in the liquid phase, not only in the water vapor phase. In this sense, the complete understanding of water drying behavior is more complete if drying curves are analyzed. In Fig. 6.13, an exemplification

Fig. 6.13 Water drying test (example).

of drying test is represented. The specimens are weighted regularly and the loss of water is quantified. Through this water loss over time, it is possible to draw the drying curves of the mortars. In these curves, different drying phases can be noticed, characterized by distinct water transport mechanisms, as illustrated in Section 2.3.3. In drying, water is transported in liquid and vapor forms [85]. The first phase is characterized by the movement of liquid water toward the surface of the specimens, followed by evaporation. The next phase or phases are characterized by the recess of the evaporation front and an increase of water vapor transport, while the liquid transport still occurs up to the evaporation front, that is it is a mixed transport. In the last phase, the water transport is characterized by water vapor diffusion only. The drying rate is the mass of water transported through the specimen per area and time units [104]. The first drying rate is analyzed with the graphic of evaporated water by time and the other rates are analyzed as a function of the squared root of time. In drying curves, the drying index is another mathematical tool that can indicate the ability of the mortars to release the amount of water contained in its pores. According to Grilo et al. [86], a lower drying index reflects a globally easier drying behavior.

Fig. 6.14 presents the drying curves of three render mortars with the same binder but with distinct volumetric ratios binder: aggregate (1:4, 1:5, and 1:6). As represented by the curves, the mortar with less cement content (mortar with a volumetric ratio of 1:6) presents higher drying rates, noticed by higher slopes of the curve tangents. This mortar also presents an area under the curves lower than the remaining ones (drying index), meaning that it dries more easily than the others. That is justified by the porosity of the mortar. The mortar with less cement content (1:6) is more porous than mortars with higher volume of cement, as mortars with a volumetric ratio of 1:5 or 1:4.

Wastes modify the render mortars' porosity and consequently their drying behavior. Some improve this behavior and others hinder drying, as will be presented. From Figs. 6.15–6.18, several drying curves examples of render mortars with wastes'

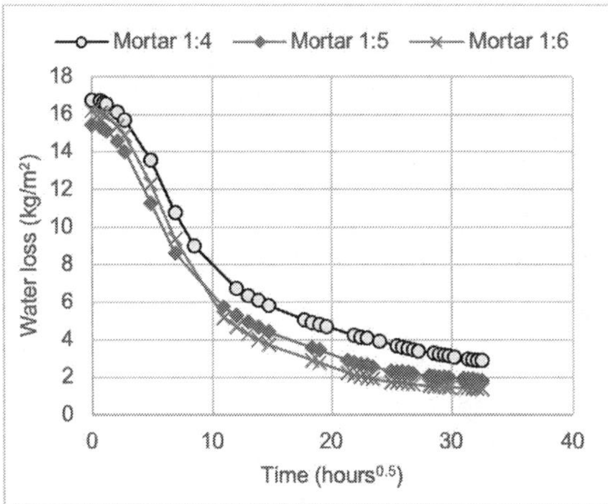

Fig. 6.14 Example of water loss curves of mortars with volumetric ratios of 1:4, 1:5, and 1:6 [1].

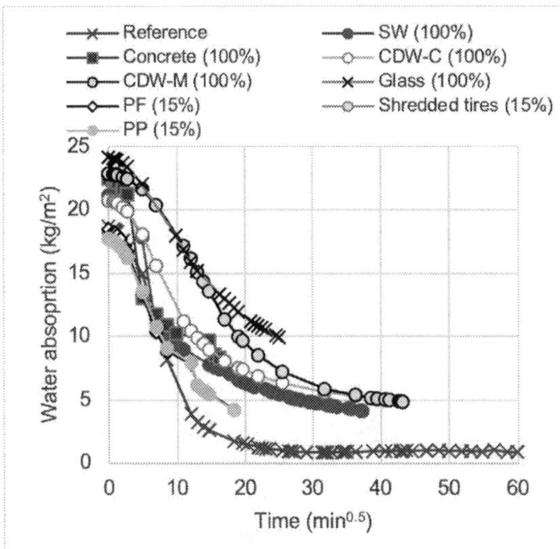

Fig. 6.15 Example of drying curves of mortars with aggregate wastes [8] [18] [137] [143] [155].

incorporation are presented. The drying rates and drying indexes from several studies that incorporated wastes in renders and analyzed their influence in drying behavior are presented in Tables 6.5–6.7. The relative rates and indexes between the modified mortars and the correspondent reference mortars are illustrated in Figs. 6.19–6.21.

Aggregate wastes incorporation in render mortars, due to their intrinsic characteristics such as porosity, shape, and roughness, give origin to more porous mortars. This higher porosity is responsible for a higher water absorption, as previously presented, but can also be responsible for faster drying mortars. For example, replacement of

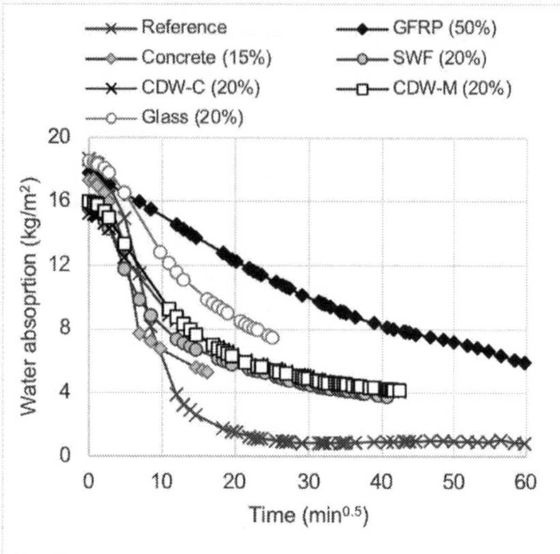

Fig. 6.16 Example of drying curves of mortars with filler wastes (fillers) [7] [51] [142] [154] [172].

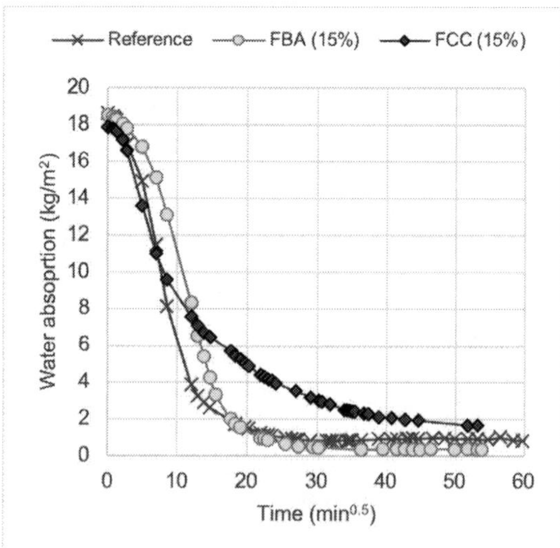

Fig. 6.17 Example of drying curves of mortars with binder wastes [215] [218].

sand by 100% sanitary ware wastes [18] results in a mortar with a similar drying rate in liquid phase, meaning that the initial drying velocity is similar. However, when the water transport starts to be a combination of liquid and vapor transport, the mortar with sanitary ware waste presents a quicker water loss, presenting a drying ratio of 2.03, which corresponds to an increment of 70% in water release speed than the

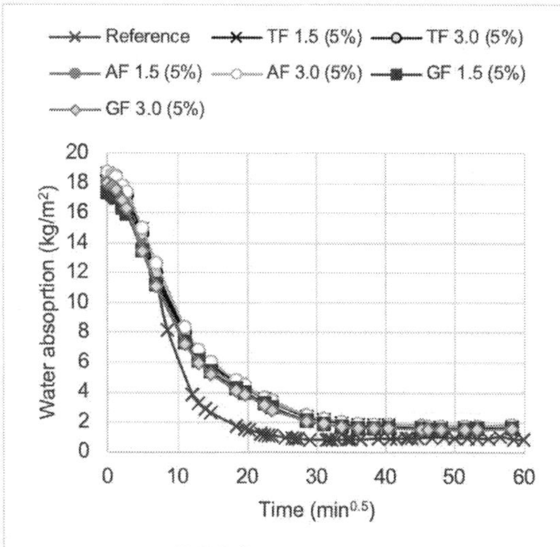

Fig. 6.18 Example of drying curves of mortars with fiber waste [218].

reference mortar [18]. However, this clear improvement of water drying behavior is not noticed for all the aggregate wastes. Glass, for example, also presents a similar rate in liquid phase (Table 6.5); however, a lower second phase ratio was noticed for these mortars (Table 6.6). Figs. 6.19 and 6.20 present the relative drying rates of the modified mortars and the reference mortars; it is noticed that there is no clear trend for aggregate wastes as far as water drying behavior is concerned. Fig. 6.21 represents the relative drying index of these modified mortars (by replacing the natural aggregates with wastes). In this index, a clearer trend can be noticed: mortars with aggregate wastes present a slower global drying behavior than the reference mortars.

Concerning filler wastes, drying rate trends are clearer. This type of wastes is responsible for lower drying rates and higher drying indexes. The replacement of the natural aggregate with thin particles (fillers) results in mortars less porous, less absorbent, and also with slower drying. For example, when sanitary ware replaces natural aggregate at 20%, incorporated as a filler [7], the modified mortar presents a similar drying rate in liquid phase (that corresponds to the first 5 h) than the reference mortar. However, when the transport of water until the surface of the specimens starts to be a combination of liquid and vapor transport (second phase), the modified mortar presents a lower drying rate than that of the reference mortar (0.96 vs 1.19 $kg/m^2 \cdot h^{1/2}$). The drying indexes of mortars with incorporation of filler wastes (Fig. 6.21) are higher than that of the reference mortar and increase with the volume of wastes incorporated.

The incorporation of fibrous materials as fiber wastes results in mortars with higher drying rates, in both water transport phases and lower drying indexes. These results mean that the incorporation of fibers facilitates the release of water in the interior of the mortar pores.

Table 6.5 Drying rate of mortars with incorporation of wastes (kg/m^2 •h) - Liquid transportation phase (first phase).

Type of incorporation	Waste	0%	5%	10%	15%	20%	50%	100%	References
Aggregate	Sanitary ware	0.14	-	-	-	0.12	0.13	0.14	Calculated from the data obtained by [18]
	Concrete	0.24	-	-	-	0.18	0.33	0.16	Calculated from the data obtained by [8]
	Construction and demolition waste (predominantly concrete)	0.12	-	-	-	0.08	0.05	0.11	Calculated from the data obtained by [155]
	Construction and demolition waste (mixture)	0.12	-	-	-	0.07	0.07	0.05	Calculated from the data obtained by [155]
	Glass	0.11	-	-	-	0.10	0.11	0.10	Calculated from the data obtained by [143]
	Plastics (PET flakes)	0.23	0.18	0.29	0.28	-	-	-	Calculated from the data obtained by [137]
	Plastics (PET pellets)	0.23	0.25	0.25	0.26	-	-	-	Calculated from the data obtained by [137]
Filler	Sanitary ware powder	0.14	-	0.14	0.15	0.17	-	-	Calculated from the data obtained by [7]
	Concrete powder	0.24	0.22	0.18	0.23	-	-	-	Calculated from the data obtained by [51]
	Construction and demolition waste powder (predominantly concrete)	0.11	-	0.12	0.16	0.14	-	-	Calculated from the data obtained by [154]
	Construction and demolition waste powder (mixture)	0.11	-	0.09	0.09	0.14	-	-	Calculated from the data obtained by [154]
	Glass powder	0.11	-	0.10	0.08	0.09	-	-	Calculated from the data obtained by [142]
	GFRP (Glass Fibre Reinforced Polymer) powder	0.10	-	0.09	0.08	0.09	0.07	-	[172]
Binder	Forest biomass ashes	0.10	0.19	0.17	0.18	-	-	-	[215]
	Fluid catalytic cracking	0.10	0.16	0.19	0.16	-	-	-	[218]
Fibre	Textile fibres (1.5 cm long)	0.10	0.20	-	-	-	-	-	[218]
	Textile fibres (3.0 cm long)	0.10	0.19	-	-	-	-	-	[218]
	Acrylic fibres (1.5 cm long)	0.10	0.21	-	-	-	-	-	[218]
	Acrylic fibres (3.0 cm long)	0.10	0.17	-	-	-	-	-	[218]
	Glass fibres (1.5 cm long)	0.10	0.19	-	-	-	-	-	[218]
	Glass fibres (3.0 cm long)	0.10	0.18	-	-	-	-	-	[218]

Table 6.6 Drying rate of mortars with incorporation of wastes ($kg/m^2 \cdot h^{1/2}$) – Mixed transportation phase (second phase).

Type of incorporation	Waste	0%	5%	10%	15%	20%	50%	100%	References
Aggregate	Sanitary ware	1.19	-	-	-	1.91	1.89	2.03	Calculated from the data obtained by [18]
	Concrete	2.13	-	-	-	2.35	2.37	2.30	Calculated from the data obtained by [8]
	Construction and demolition waste (predominantly concrete)	0.78	-	-	-	0.90	0.76	1.04	Calculated from the data obtained by [155]
	Construction and demolition waste (mixture)	0.78	-	-	-	0.66	0.72	0.85	Calculated from the data obtained by [155]
	Glass	0.93	-	-	-	0.76	0.82	0.81	Calculated from the data obtained by [143]
	Plastics (PET flakes)	1.43	1.43	1.37	1.39	-	-	-	Calculated from the data obtained by [137]
	Plastics (PET pellets)	1.43	1.18	1.30	1.22	-	-	-	Calculated from the data obtained by [137]
Filler	Sanitary ware powder	1.19	-	0.96	0.94	0.96	-	-	Calculated from the data obtained by [7]
	Concrete powder	2.13	1.88	1.81	1.81	-	-	-	Calculated from the data obtained by [51]
	Construction and demolition waste powder (predominantly concrete)	0.81	-	0.73	0.61	0.59	-	-	Calculated from the data obtained by [154]
	Construction and demolition waste powder (mixture)	0.81	-	0.59	0.61	0.66	-	-	Calculated from the data obtained by [154]
	Glass powder	0.93	-	0.77	0.71	0.69	-	-	Calculated from the data obtained by [142]
	GFRP (Glass Fibre Reinforced Polymer) powder	0.76	-	0.18	0.17	0.17	0.23	-	[172]
Binder	Forest biomass ashes	0.76	0.79	0.96	0.92	-	-	-	[215]
	Fluid catalytic cracking	0.76	0.81	1.24	1.24	-	-	-	[218]
Fibre	Textile fibres (1.5 cm long)	0.76	0.86	-	-	-	-	-	[218]
	Textile fibres (3.0 cm long)	0.76	0.9	-	-	-	-	-	[218]
	Acrylic fibres (1.5 cm long)	0.76	0.84	-	-	-	-	-	[218]
	Acrylic fibres (3.0 cm long)	0.76	0.94	-	-	-	-	-	[218]
	Glass fibres (1.5 cm long)	0.76	0.84	-	-	-	-	-	[218]
	Glass fibres (3.0 cm long)	0.76	0.85	-	-	-	-	-	[218]

Table 6.7 Drying index of mortars with incorporation of wastes.

Type of incorporation	Waste	0%	5%	10%	15%	20%	50%	100%	References
Aggregate	Sanitary ware	0.24	-	-	-	0.23	0.26	0.28	Calculated from the data obtained by [18]
	Concrete	0.31	-	-	-	0.36	0.39	0.46	Calculated from the data obtained by [8]
	Construction and demolition waste (predominantly concrete)	0.28	-	-	-	0.28	0.30	0.31	Calculated from the data obtained by [155]
	Construction and demolition waste (mixture)	0.28	-	-	-	0.32	0.32	0.32	Calculated from the data obtained by [155]
	Glass	0.43	-	-	-	0.52	0.51	0.55	Calculated from the data obtained by [143]
	Plastics (PET flakes)	0.31	0.33	0.36	0.37	-	-	-	Calculated from the data obtained by [137]
	Plastics (PET pellets)	0.31	0.39	0.37	0.39	-	-	-	Calculated from the data obtained by [137]
Filler	Sanitary ware powder	0.24	-	0.30	0.33	0.33	-	-	Calculated from the data obtained by [7]
	Concrete powder	0.31	0.35	0.37	0.37	-	-	-	Calculated from the data obtained by [51]
	Construction and demolition waste powder (predominantly concrete)	0.28	-	0.32	0.34	0.36	-	-	Calculated from the data obtained by [154]
	Construction and demolition waste powder (mixture)	0.28	-	0.36	0.38	0.34	-	-	Calculated from the data obtained by [154]
	Glass powder	0.43	-	0.50	0.52	0.53	-	-	Calculated from the data obtained by [142]
	GFRP (Glass Fibre Reinforced Polymer) powder	0.25	-	0.44	0.46	0.52	0.54	-	[172]
Binder	Forest biomass ashes	0.25	0.14	0.16	0.16	-	-	-	[215]
	Fluid catalytic cracking	0.25	0.22	0.22	0.21	-	-	-	[218]
Fibre	Textile fibres (1.5 cm long)	0.25	0.16	-	-	-	-	-	[218]
	Textile fibres (3.0 cm long)	0.25	0.16	-	-	-	-	-	[218]
	Acrylic fibres (1.5 cm long)	0.25	0.16	-	-	-	-	-	[218]
	Acrylic fibres (3.0 cm long)	0.25	0.17	-	-	-	-	-	[218]
	Glass fibres (1.5 cm long)	0.25	0.16	-	-	-	-	-	[218]
	Glass fibres (3.0 cm long)	0.25	0.15	-	-	-	-	-	[218]

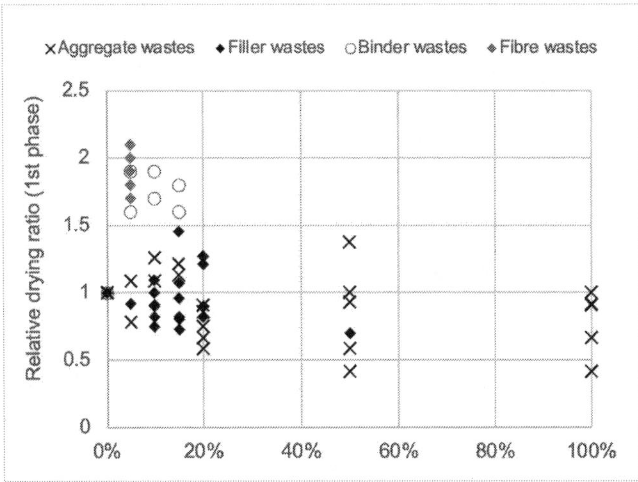

Fig. 6.19 Relative drying rate (first phase) of mortars with aggregate, filler, binder, or fiber wastes [7] [8] [18] [51] [137] [142] [143] [154] [155] [172] [215] [218].

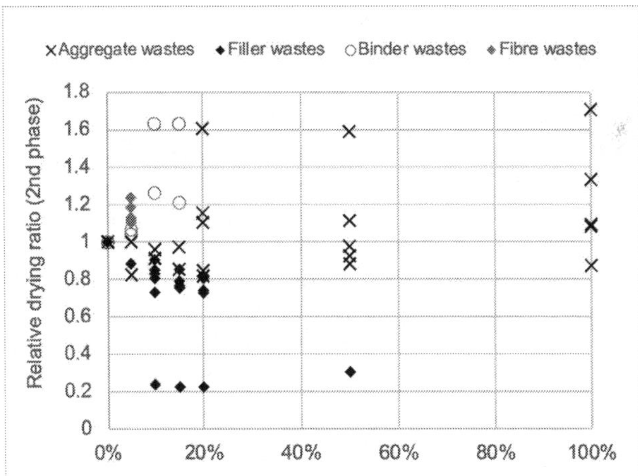

Fig. 6.20 Relative drying rate (second phase) of mortars with aggregate, filler, binder, or fiber wastes [7] [8] [18] [51] [137] [142] [143] [154] [155] [172] [215] [218].

Similarly, the replacement of the binder with wastes also results in mortars with a higher drying rate. In this sense, these types of wastes (fibers or binder wastes) prevent the water, after reaching the mortars, from remaining there for a long period of time. In this sense, fibers and binders are types of wastes that clearly improve the drying behavior of the mortars. Aggregate wastes, in some cases, also improve this mortar's property. Filler wastes, on the other side, are in most cases responsible for a more difficult water release process than the remaining wastes.

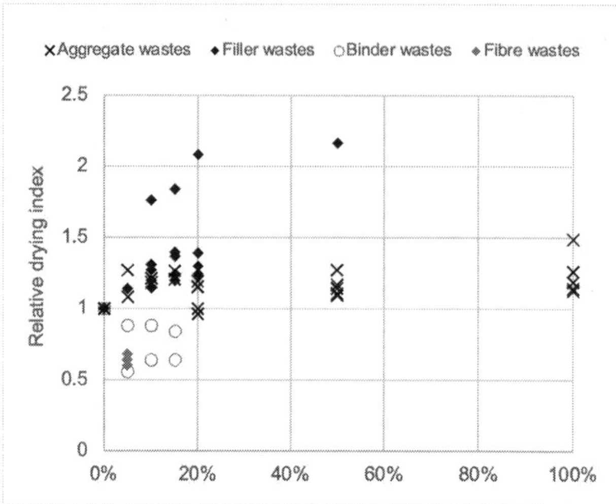

Fig. 6.21 Relative drying index of mortars with aggregate, filler, binder, or fiber wastes. References: [7] [8] [18] [51] [137] [142] [143] [154] [155] [172] [215] [218].

6.3 Chapter conclusions

Renders and plasters are mortars that can be applied on a substrate, functioning as a sacrificial layer to increase the service life of the substrate. Render mortars are submitted to the most aggressive external actions, such as wind, rain, storms, or others. Render mortars should be able to prevent the water from migrating to the substrate, avoiding their degradation, and consequently incrementing their lifecycle. The render mortars water can absorb water by capillarity or with pressure. The pressure can be associated to wind, which blows on the surface and creates some pressure, or by other type of pressure as a vertical drainage of water that, due to gravity forces, originates pressure on the mortars. In this chapter, the influence of cement content and wastes incorporation on water absorption was analyzed. It was concluded that different wastes present different water behaviors. The wastes were divided in four groups: aggregates, fillers, fibers, and binders. In general, it was found that fine wastes, incorporated in the mortars as fillers, make them less permeable to liquid water and to water vapor than the reference mortars. Aggregates, fibers, and binder wastes, on the other hand, and in general, absorb a higher amount of liquid water, being more permeable. In this regard, it is also mentioned that, whenever the render mortars are in contact with water, they should release the amount absorbed whenever the external conditions allow it. This release of water, over time, was also evaluated by some authors, and some examples of these curves in the modified mortars were also discussed in this chapter. The wastes that absorb higher amounts of water are usually those that also release it easier. In this sense, fiber and binder wastes are the ones that dry faster.

Durability

7

During their lifetime, render mortars are subjected to several external actions that can reduce their initial performance. Mortars reach their end of lifetime when their characteristics do not meet their functional requirements. Durability is the capacity of the mortars to maintain functional requirements for the longest time possible. A durable mortar can last for centuries as proved by many examples from Antiquity [78].

Durability does not depend only on the mortars' characteristics. It also depends on the compatibility between the mortar and the substrate, on the surrounding environment and whether it was well designed for it, on the efficiency the application, and on other occasional or fortuitous actions that can occur. Thus, durability is a characteristic hard to predict as it depends on many variables.

There are several anomalies that can reduce render mortars' durability such as the lack of adherence between the mortars and their substrate or cracking. This last anomaly can be due to the movements of the substrate, due to ice or salt formation, due to chemical reactions (as alkalis-silica), or due to human blows or to other actions. Cracking is responsible for the reduction of mortars' mechanical strengths and it is an open path for water to penetrate into deeper layers of the render, being able to achieve the substrate. Thus, cracking can be responsible for the reduction of mortars' lifetime.

Although it is difficult to predict, render mortars' durability can be evaluated by some tests, namely by ageing tests. Ageing can be either natural or artificial. The former intends to represent natural ageing, where the mortars are submitted to different actions, such as temperature variations and water exposition, usually by a resource to cycles of similar but more severe conditions. Artificial ageing is an essential process to evaluate the durability of the mortars when the natural process is not possible or viable due to its duration. However, artificial ageing does not include all the degradation mechanisms and the effect of real time is sometimes difficult to simulate. In this chapter, the influence of wastes on mortars' durability is discussed.

7.1 Compatibility with substrate

Render mortars are applied over a substrate, mainly aiming at protecting it from the external actions that the external layer of the buildings is submitted to during their lifetime. This protection prolongs the substrate's lifetime. External coatings also have aesthetic functions, thus renders should also meet adequate aesthetic requirements.

A proper compatibility between the render mortars and their substrate will significantly influence their performance and durability. This compatibility includes chemical, mechanical, and physical compatibilities.

The render and their substrate should be chemically compatible. The render should not contain compounds that react with the substrate and give origin to anomalies. For example, if the mortars contain salts, they can migrate to the substrate and be

Eco-efficient Rendering Mortars. DOI: 10.1016/C2018-0-04147-X responsible for degradation of the element that the mortar was supposed to protect. Chemical compatibility between these two elements is fundamental to prolong the lifetime of the substrate.

It is also essential to achieve mechanical compatibility between these two elements. The render mortar should be able to adhere to the substrate. It should fulfill the irregularities of the substrate and penetrate into the substrate's superficial pores, creating a proper connection to it. There are some physical actions, such as thermal or humidity variations, which can generate stresses in the render/substrate interface that can originate the mortars' detachment in zones where this interface strength is lower. In addition, for suitable mechanical compatibility, the render should also have mechanical strengths adapted to the substrate [77]. When a substrate presents low mechanical strength, the render should not be overly hard or brittle to avoid transmitting stresses to the substrate, damaging it [77].

In addition to a chemical and a mechanical compatibility, physical compatibility is necessary. For this purpose, the mortars' deformability is important. A similar deformability between the mortars and their substrate is essential to prolong the mortars' service life and consequently the substrate's durability. The mortar and its substrate should function as a single element that moves, dilates, or contracts similarly. This similarity reduces the stresses in the interface of the two materials, decreasing cracking's probability. Thus, the compatibility in terms of deformability reduces renders' cracking. The physical compatibility also includes other characteristics. For example, rising damp is a common effect that happens in old, very porous, walls. The renders applied over very porous walls should be able to release the amount of water absorbed by the wall, in this sense they should be very permeable to water vapor. For common contemporary walls, it is still important to assure relatively high water vapor permeability of the render to allow the release of water vapor produced inside the buildings and quick drying of the several layers of the wall where liquid water infiltrations can occur. A low permeability render is responsible for retaining water for a longer time in the interior of the mortars' pores accelerating the walls' degradation.

7.2 Ageing—how can it affect mortars durability?

7.2.1 Deformability over time

Mortars deformability can be modified over time, being an indicator of durability. Deformability can increase or decrease depending on the reactions that occur in the mortars' matrix.

Cracking or internal microcracking are anomalies that can reduce the mortars modulus of elasticity (ME). The matrix becomes less compact with cracking, which is responsible for this decrease. Consequently, the matrix is weaker, decreasing its mechanical strengths. This cracking or internal microcracking can occur during ageing, modifying mortars' deformability.

Render mortars' cracking can be due to several actions, as discussed in this book. Shrinkage, thermal or humidity gradients, blows, or movements of the substrate can lead to cracks. The gravity and extension of these cracks are related to the initial mortars' ME. For any of these actions, a more deformable mortar will present a better behavior than a more brittle one. A deformable mortar submitted to extreme actions can also crack. However, the cracks pattern will be less severe than in a more brittle mortar: cracks will present a lower width, lower length, and lower depth. Independently of the initial deformability, whenever a mortar cracks, its ME is reduced.

Internal microcracking can also occur due to several events. Ice or salts formation causes mechanical stresses that are imposed by the water in a solid state or by salts that are created in the mortars' pores and can result in internal microcracking. Matrix shrinkage is another action that can result in internal microcracking, namely in the interface aggregate/matrix. With shrinkage the cement matrix contracts decreasing the volume of voids but simultaneously creates stresses in the interface transition zone (ITZ) of the aggregates/matrix, which can lead to degradation of this interface resulting in internal microcracks [75]. This pattern is responsible for a reduction of the ME in mortars.

The late hydration of the binder is a common reaction that occurs during ageing. This late hydration occurs when the constituents of the cement that did not react previously, react later in the presence of water giving origin to new stable cement components. The compounds formed will occupy more space than the previous unhydrated compounds increasing the mortars compactness. Therefore, these reactions are responsible for increments in strength over time and consequently an increase of the ME is noticed.

In fact, a modification of the mortars' deformability can occur over time. This modification can be responsible for a change, not only in the mortar's ME but in other characteristics. The incorporation of wastes can also be responsible for a modification of these properties over time. Next, two different sets of tests to evaluate the ME modifications over time are presented.

In the first set, the results (over time) of the ME by the frequency of resonance test of several mortars are presented. In those tests, the mortars (prismatic specimens 40 × 40 × 160 mm) were maintained in controlled temperature and humidity during the period under analysis. In the second test, mortars' ME is evaluated by ultrasound pulse velocity before and after artificial ageing. In this analysis, mortars were applied over a ceramic substrate and were submitted to an artificial ageing.

In Fig. 7.1, examples of ME of mortars with aggregate, filler, binder, or fiber wastes are presented (first set of tests). The values were obtained over time for mortars that were maintained in controlled conditions of temperature and humidity (20 ± 2 °C and 65 ± 5%) until testing. In the figure, the ME over time presented is related with the one obtained at 28 days. Thus, it is possible to observe that the ME of most mortars (with wastes' incorporation) decreased over time. Only a few exceptions were noticed for increments of this modulus, as will be shown next. The ME test was evaluated in prismatic specimens (Fig. 7.2). The reasons for the modification of the ME over time are described later.

In those prisms, the mortar can shrink freely. With free shrinkage, cracking does not occur with the same magnitude as in restrained conditions, which is the case

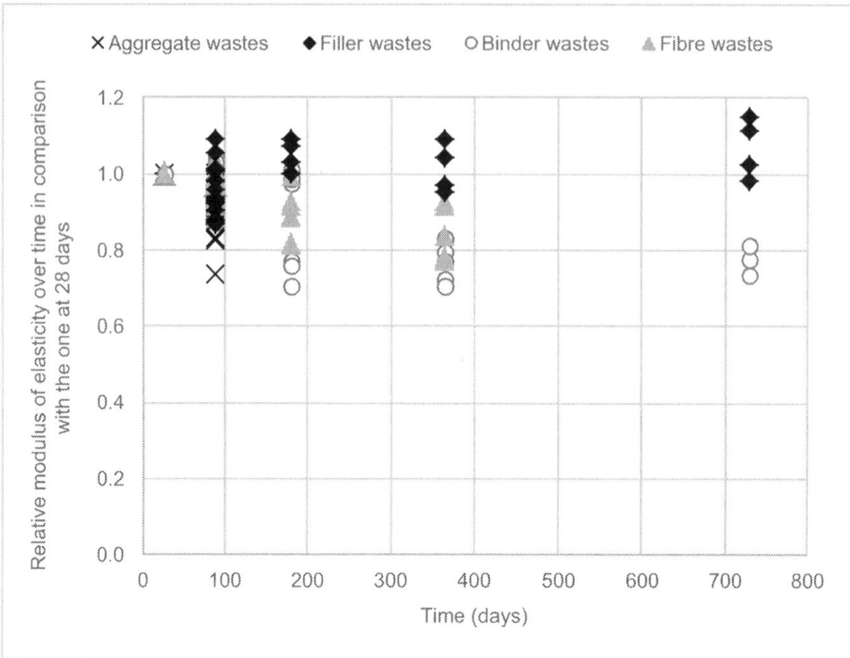

Fig. 7.1 Relative modulus of elasticity obtained over time for mortars with aggregate, filler, binder, or fiber wastes in comparison with the one obtained at 28 days [7] [18] [126] [137] [142] [143] [148] [154] [155] [172] [215] [218] [223].

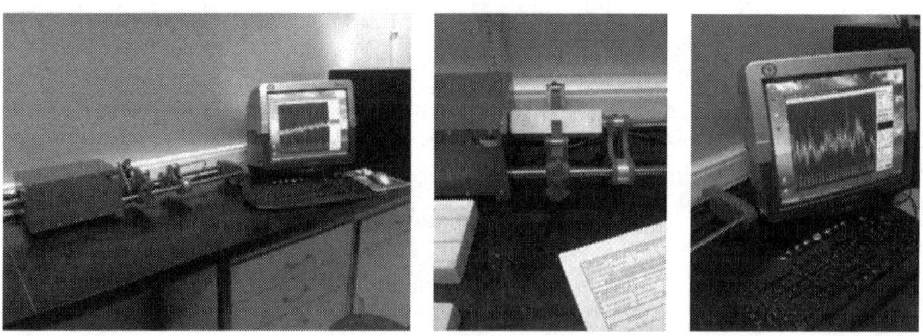

Fig. 7.2 Modulus of elasticity test representation.

of the second set of tests presented next, where the mortars were applied over a ceramic brick and submitted to ageing. In the prisms, on the one hand, shrinkage (without restriction) can increase the compactness of the mortars and consequently their ME. On the other hand, a high shrinkage could create stresses between the aggregates and their matrix, and be responsible for internal microcracking at the interface, which results in a reduction of the ME [75]. With a controlled relative humidity and without direct contact with water, the specimens were not submitted to the late hydration as described above, which could be responsible for an increase of the ME, as discussed previously. However, they can suffer other chemical reactions, for example carbonation, which can modify this modulus. The prisms had dimensions of 40 mm × 40 mm × 160 mm, and six faces in contact with air, which is favorable for the carbonation reaction and consequently increments their compacity and their ME.

In Tables 7.1–7.5, the exact values of ME, over time, of the mortars illustrated in Fig. 7.1 are presented.

In Fig. 7.1, it was noticed that most of the mortars with wastes' incorporation decreased their ME from 28 days until longer ages, with some exceptions for filler wastes. A filler waste that did not follow this trend was the glass fiber reinforced polymer (GFRP) waste, for different percentages of incorporation. The remaining mortars with wastes presented in general reductions of this property. The reduction in the ME has implications on the mortars' mechanical behavior as well as on the water behavior. The reduction of the ME is related to internal microcracking; the mortars' shrinkage was probably responsible for creating stresses in the ITZ giving origin to internal microcracking and consequently decreasing the ME. In mortars with filler wastes, the reduction of the ME resulted in decreases in mechanical strength. For example, the mortar with 20% of sanitary ware (powder) presented from 28 days to 90 days a reduction of the ME of 7%, which corresponded to a reduction from 18.6 GPa to 17.3 GPa [7]. The reduction of the ME over time was followed by a reduction of compressive strength of 34%, from 14.92 MPa to 9.55 MPa (see Section 5.1.1.). Another example is forest biomass ashes (FBA) that reduced the ME of the mortar from 28 days to 730 days, from 8.1 GPa to 6.6 GPa [215], which corresponds to reductions of 19% of this modulus and 23% of the compressive strength, from 4.30 MPa to 3.32 MPa. In the case of FBA, the reduction of the ME of the mortar is higher than the one caused by sanitary ware; however, the decrease of strength of the sanitary ware mortar is higher. In fact, it can be noticed that, for mortars with higher ME, namely the filler waste mortars, the reduction of the ME represents higher reductions of mechanical strength. In fact, the filler waste mortars are more brittle and whenever they crack the reduction of strength is more noticeable.

To evaluate the ME after ageing, another set of tests is presented. In this case, the mortars were applied over a ceramic substrate (2 cm of mortar thickness) and their ME was measured by the ultrasound pulse velocity test before and after artificial ageing. In this ageing process, the mortars were submitted to several cycles of humidity-freeze and heat-freeze according to European Standard EN 1015-21 [90].

In Fig. 7.3, the specimens used by Farinha [218] to evaluate the ME before and after artificial ageing are illustrated. Three alinements were evaluated, and the curves of distance/time were drawn. From the tangent to the curves, ultrasound pulse velocity can be determined. The calculation of this dynamic ME is given by Eq. (7.1), where K is calculated from Eq. (7.2), ν is the Poisson coefficient that was considered equal to 0.2, V the ultrasound pulse velocity obtained and ρ the bulk density of the mortar.

Table 7.1 Modulus of elasticity over time of mortars with filler wastes' incorporation obtained by several authors—first set of tests.

Type of waste	Specimens type	Incorporation ratio (%)	Modulus of elasticity (GPa)				References
			28 days	60 days	90 days	150 days	
Red ceramics	Prismatic samples (40 × 40 × 160 mm)	50	-	8.7	-	8.3	[221]
Sanitary ware		20	11.9	-	11.5	-	[18]
		50	11.6	-	11.1	-	
		100	8.2	-	8.1	-	
CDW (predominantly concrete)		20	10.2	-	8.9	-	[155]
		50	8.9	-	7.4	-	
		100	8.3	-	6.1	-	
CDW		20	10.8	-	9.8	-	[155]
		50	6.4	-	5.8	-	
		100	4.5	-	4.0	-	
Glass		20	10.2	-	9.3	-	[143]
		50	11.0	-	10.0	-	
		100	9.0	-	8.3	-	
Shredded tires		5	7.0	-	6.8	-	[148]
		10	5.7	-	5.7	-	
		15	5.4	-	4.8	-	
PET flakes		5	6.3	-	5.8	-	[137]
		10	5.2	-	4.3	-	
		15	4.5	-	4.1	-	
PET pellets		5	5.7	-	5.3	-	[137]
		10	5.4	-	4.7	-	
		15	5.0	-	4.6	-	

Table 7.2 Modulus of elasticity over time of mortars with filler wastes' incorporation obtained by several authors—first set of tests.

Type of waste	Incorporation ratio (%)	Specimens type	Modulus of elasticity (GPa)							References
			28 days	60 days	90 days	150 days	180 days	365 days	730 days	
Red ceramics	10	Prismatic samples (40 × 40 × 160 mm)	-	12.2	-	15.6	-	-	-	[49]
Sanitary ware	10		15.7	-	14.6	-	-	-	-	[7]
	15		17.4	-	16.1	-	-	-	-	
	20		18.6	-	17.3	-	-	-	-	
CDW (predominantly concrete)	10		14.1	-	12.2	-	-	-	-	[154]
	15		15.4	-	14.3	-	-	-	-	
	20		15.2	-	15.3	-	-	-	-	
CDW	10		13.3	-	12.0	-	-	-	-	[154]
	15		11.9	-	11.7	-	-	-	-	
	20		10.6	-	9.8	-	-	-	-	
Glass	10		9.9	-	8.7	-	-	-	-	[142]
	15		11.0	-	10.1	-	-	-	-	
	20		12.5	-	10.9	-	-	-	-	
GFRP	10		9.6	-	9.2	-	9.6	9.3	9.4	[172]
	15		9.7	-	10.2	-	10.4	10.1	9.9	
	20		10.2	-	11.1	-	11.1	9.7	11.7	
	50		9.9	-	10.0	-	10.2	10.8	11.0	

Table 7.3 Modulus of elasticity over time of mortars with binder wastes' incorporation obtained by several authors—first set of tests.

Type of wastes	Incorporation ratio (%)	Specimens type	Modulus of elasticity (GPa)					References
			28 days	90 days	180 days	365 days	730 days	
Forest biomass ashes	5	Prismatic samples (40 × 40 × 160 mm)	9.9	9.0	7.6	7.6	7.7	[215]
	10		9.4	8.2	7.1	6.8	6.9	
	15		8.1	7.4	5.7	5.7	6.6	
Fluid Catalytic Cracking (FCC)	5		8.2	8.6	8.3	6.8	-	[218]
	10		7.3	7.5	7.1	5.8	-	
	15		7.0	6.7	6.9	5.4	-	

Table 7.4 Modulus of elasticity over time of mortars with fiber wastes' incorporation obtained by several authors—first set of tests.

Type of fiber	Addition percentage	Characteristics (cm)	Specimens type	Modulus of elasticity (GPa)				References
				28 days	90 days	180 days	365 days	
Sheep's wool	10%	L: 1.5	Prismatic samples (40 × 40 × 160 mm)	16.0	15.2	14.8	-	[79]
		L: 3.0		12.8	11.7	11.4	-	
	20%	L: 1.5		15.3	14.3	13.6	-	
		L: 3.0		16.9	17.5	16.8	-	
Textile	5%	L: 1.5		8.6	7.6	7.0	6.6	[218]
		L: 3.0		8.7	7.8	7.1	6.8	
	10%	L: 1.5		12.7	12.3	-	-	[223]
		L: 3.0		13.1	12.8	-	-	
	20%	L: 1.5		10.9	11.0	-	-	
		L: 3.0		11.6	11.5	-	-	
Acrylic	5%	L: 1.5		8.4	8.3	7.5	7.8	[218]
		L: 3.0		8.4	7.9	7.7	7.0	
Glass	5%	L: 1.5		9.4	8.8	8.6	8.6	[218]
		L: 3.0		9.4	8.3	8.3	7.3	

Table 7.5 Modulus of elasticity before and after artificial ageing by several authors—second set of tests.

Type of waste	Waste	Incorporation ratio (%)	Specimens type	Modulus of elasticity (GPa)		Comparison of the modulus of elasticity before and after ageing (%)	References
				Before ageing	After ageing		
Aggregate	Sanitary ware	100	Mortars applied over a ceramic substrate (2 cm layer thickness)	10.4 ± 0.8	11.1 ± 1.0	+7	[218]
		50		3.3 ± 0.2	5.9 ± 1.0	+8	[18]
	Glass	20		4.0	3.3	-18	[143]
		50		3.6	3.6	Equal	[143]
		100		2.4	2.8	+17	[143]
Filler	GFRP	50		8.9 ± 0.8	9.7 ± 2.2	+10	[218]
Binder	Forest biomass ashes	15		11.3 ± 0.6	10.0 ± 1.8	-12	[218]
Fiber	Textile fibers	5		10.0 ± 0.3	11.2 ± 1.1	+12	[218]

Table 7.6 Adherence strength and fracture type of mortars with wastes' incorporation after ageing from several authors.

Type of incorporation	Waste	0%	5%	10%	15%	20%	50%	100%	References
Aggregates	Red ceramics	0.68	-	-	-	-	0.80	-	[221]
		B					A		
	Sanitary ware	0.71	-	-	-	-	0.83	-	[18]
		C					B/C		
		0.51	-	-	-	-	-	0.37	[218]
		B						B	
	Glass	1.08	-	-	-	0.89	0.72	0.70	[143]
		B				B	B	B	
	Shredded tires	0.66	-	-	0.66	-	-	-	[148]
		B			B				
	PET (Polyethylene terephthalate) flakes	0.29	-	-	0.14	-	-	-	[137]
		-			-				
	PET (Polyethylene terephthalate) pellets	0.29	-	-	0.23	-	-	-	[137]
		-			-				
Filler	Red ceramics powder	0.68	-	0.85	-	-	-	-	[49]
		B		A					
	Sanitary ware powder	0.71	-	-	-	0.77	-	-	[7]
		A				A			
	CDW (predominantly concrete) powder	0.55	-	-	-	0.66	-	-	[154]
		B				A			
	CDW powder	0.55	-	-	0.64	-	-	-	[154]
		B			A				
	GFRP powder	0.51	-	-	-	-	0.28	-	[218]
		B					A		
Binder	Forest biomass ashes	0.51	-	-	0.32	-	-	-	[218]
		B			B				
Fibre	Textile fibres	0.51	0.43	-	-	-	-	-	[218]
		B	B						

Note: Fracture patterns: A—Adhesive fracture (in the interface mortar/substrate); Fracture pattern B—Cohesive fracture (in the mortar); Fracture pattern C—Cohesive fracture (in the substrate).

Table 7.7 Permeability to water under pressure of mortars with incorporation of wastes (ml/cm^2).

Type of incorporation	Waste	0%	5%	10%	15%	20%	50%	100%	References
Aggregate	Red ceramics	1.97	-	-	-	-	1.85	-	[221]
	Sanitary ware	4.30	-	-	-	-	6.10	-	[18]
		11.64	-	-	-	-	-	11.37	[218]
	Construction and demolition waste (predominantly concrete)	3.43	-	-	-	-	-	5.15	[155]
	Construction and demolition waste (mixture)	3.43	-	-	-	4.60	-	-	[155]
	Glass	1.49	-	-	-	2.12	2.55	3.03	[143]
Filler	Red ceramics powder	1.97	-	1.81	-	-	-	-	[49]
	Sanitary ware powder	4.33	-	-	-	0.77	-	-	[7]
	Construction and demolition waste powder (predominantly concrete)	3.43	-	-	-	2.09	-	-	[154]
	Construction and demolition waste powder (mixture)	3.43	-	-	2.57	-	-	-	[154]
	GFRP powder	11.64	-	-		-	4.96	-	[218]
Binder	Forest biomass ashes	11.64	-	-	13.71	-	-	-	[218]
Fibre	Textile fibres (1.5 cm long)	11.64	12.54	-	-	-	-	-	[218]

Table 7.8 Esthetical appearance of blended mortars with wastes' incorporation after artificial ageing evaluated by several authors.

Type of incorporation	Waste	Incorporation ratio (in volume) (%)	Appearance after ageing		References
			Cracks	Leaching	
Aggregates	Sanitary ware	50	No	Yes	[18]
		100	No	Yes	[218]
	CDW (predominantly concrete)	100	No	-	[155]
	CDW	20	No	-	[155]
	Glass	20	No	-	[143]
		50	No	-	
		100	No	-	
	PET flakes	15	No	Yes	[137]
	PET pellets	15	No	Yes	[137]
Filler	Sanitary ware	20	No	Yes	[7]
	CDW (predominantly concrete)	20	No	Yes	[154]
	CDW	15	No	Yes	[154]
	Glass	20	No	-	[142]
	Glass fiber reinforced polymer	50	Yes	Yes	[218]
Binder	Forest biomass ashes	15	No	Yes	[218]
Fiber	Textile fibers (1.5 cm long)	5	No	Yes	[218]

$$ME_{US} = K \times V^2 \times \rho \times 10^{-6} \tag{7.1}$$

$$K = \frac{(1+\nu)\times(1-2\nu)}{(1-\nu)} \approx 0.9 \tag{7.2}$$

In Table 7.4, results of ME before and after ageing obtained by some authors with this technique are summarized. In the eight mortars with wastes' incorporation that were tested, five achieved a higher ME after ageing, one presented an equal value, and only two had their ME decreased. The highest variations were noticed for the glass waste incorporation: the replacement of sand at 20% resulted in a decrease of 18% of ME (after ageing) and the replacement at 100%

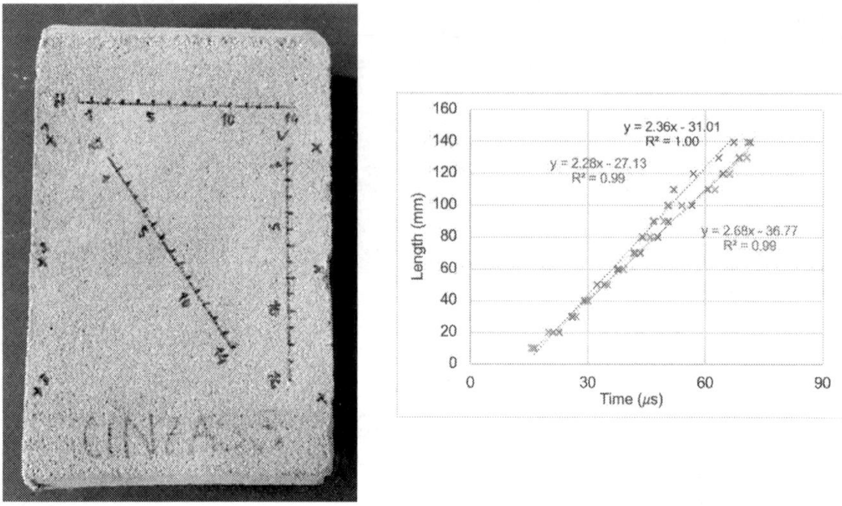

Fig. 7.3 Illustration of a modulus of elasticity test specimen and results of the ultrasound pulse velocity test [218].

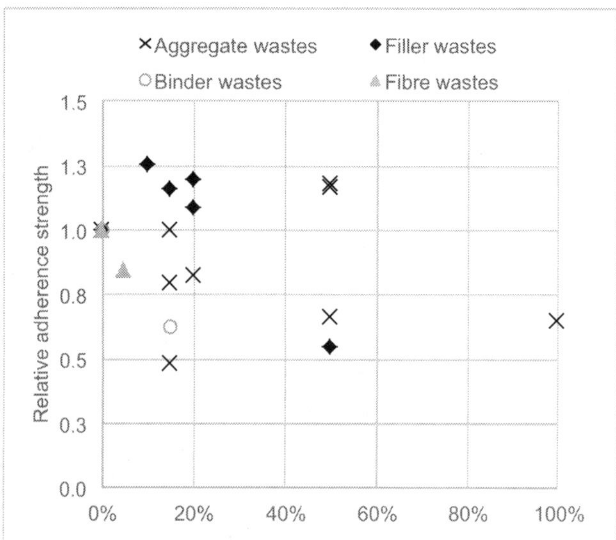

Fig. 7.4 Relative adherence strengths of mortars with aggregate, filler, binder, or fiber wastes after ageing [7] [18] [49] [137] [143] [148] [154] [218] [221].

and increment of 17% (after ageing). The increments noticed by the authors are probably due to the late hydration of the cement compounds in the humidity-freeze cycles. Besides the mortar with 20% of glass waste, the mortar with FBA replacing cement also presented an ME 12% lower after ageing than the value obtained before it. The reduction of the ME can be due to internal microcracking due to thermal or humidity variations that occurred in the heat-freezing cycles or in the humidity-freezing cycles, respectively. There were no visible cracks on the surface of the specimens, with the exception of the GFRP waste, as illustrated in Section 7.2.1.

7.2.2 Adherence before and after ageing

Adherence strength is a property that can be modified over time, due to the mortars' ageing. This property is one of the most important renders' mortars functional requirements. Without proper adherence the mortars detach from the substrate leading to the need of render's replacement and therefore to the end of their life cycle. To prolong the mortars' service life, good adherence should be maintained for the longest time possible.

There are some events that can affect the mortars' adherence. Differential deformations between the mortars and the substrates due to thermal and humidity variations create stresses that can exceed the resistance of the interface mortar/substrate. Mortars' shrinkage can also increment the stresses in the interface. The encapsulation of wastes in render mortars, replacing the usual constituents, can modify these differential tensions and consequently influence the behavior of the mortars, concerning the adherence strength, over time.

Following some examples of wastes' incorporation in render mortars and their influence on the adherence strength after ageing are presented. In all the examples, the render mortars (with wastes' incorporation) were submitted to an artificial ageing. This process consisted of eight heating-freezing cycles followed by eight wetting-freeze cycles [7] [18] [49] [137] [143] [148] [154] [218] [221].

In Table 7.6, some examples of aggregate, filler, binder, and fiber wastes and their adherence strength after artificial ageing are presented. A comparison between the value obtained by these blended mortars and their correspondent reference (REF) mortar is illustrated in Fig. 7.4.

Concerning the aggregate wastes, it is noticed by the captions that the adherence strength is lower in those blended mortars, when compared with the REF mortar, with exception of the red ceramics and sanitary ware mortars when both wastes replaced the natural aggregate at 50% (in volume). The same trend was noticed before ageing (see Chapter 5). Most of the mortars with aggregate wastes and the corresponding REF mortars presented a fracture type B or C, meaning they were cohesive in the mortar or in the substrate (in this case, bricks were used as substrates). These fracture types indicate that the adherence in the interface is higher than the value obtained. The mortar or the ceramic substrate tensile strength conditions the value obtained. In Fig. 7.5, the fracture pattern before and after ageing of a mortar with sanitary ware waste, replacing the natural aggregate by 100%, is illustrated.

Filler wastes presented, after ageing, higher adherence strengths than the corresponding REF mortars, with the exception of the mortar with GFRP powder waste. As presented next, this waste worsened the mortars' characteristics after ageing leading to an abrupt reduction of the properties. The remaining filler wastes

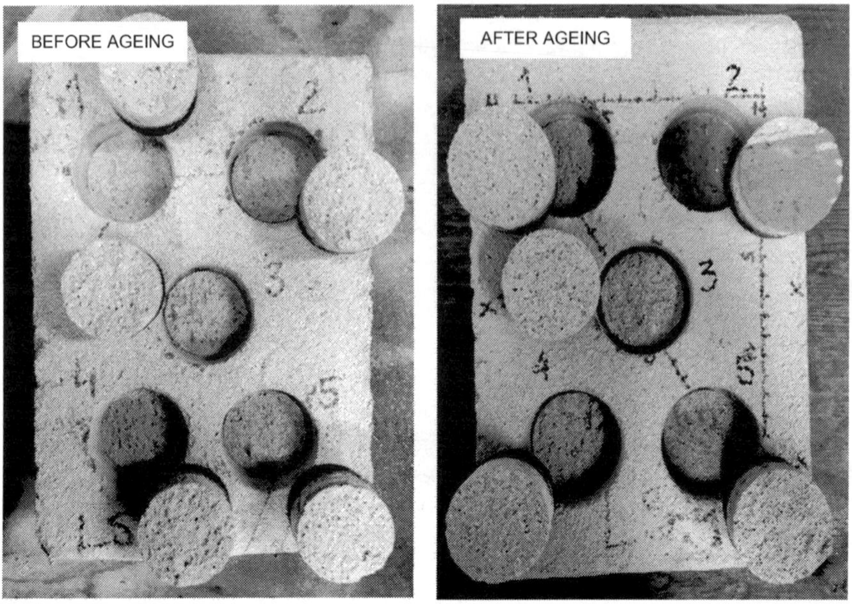

Fig. 7.5 Adherence strength fracture pattern of a mortar with sanitary ware waste (100%)—before and after ageing [218].

mortars presented strengths higher than the ones presented by the REF mortar. Contrary to the aggregate wastes, filler wastes mortars presented a fracture type A corresponding to an adhesive fracture and therefore in the interface between the mortar and the substrate. In Figs. 7.6 and 7.7, two examples of mortars with filler wastes' incorporation and the corresponding failure patterns before and after ageing are illustrated.

An example of a mortar with binder waste is also illustrated in Fig. 7.8. As noticed, the FBA from the research by Farinha [218] presented an adherence strength of 0.32 MPa, corresponding to a reduction of 37% in comparison with the one obtained by the REF mortar. This value is influenced by the mortar's mechanical strength (fracture type B), being responsible for the reduction of the adherence strength test value. Nevertheless, the interface mortar/substrate strength presented a value higher than the one obtained by the test. In Fig. 7.8, the fracture pattern of this mortar is illustrated.

An example of textile fibers reinforcing a render mortar and their influence on adherence strength is also presented in the captions. The fibers addition leads to a reduction of adherence strength of about 16% (0.43 MPa vs 0.51 MPa) in comparison with its REF mortar. Similarly to the binder waste, this mortar presented a fracture type B, cohesive in the mortar. This fracture means that the mechanical strength of the mortar directly influenced the value obtained in the test and the value of adherence strength in the interface mortar/substrate is higher than the value obtained. An illustration of this fracture pattern is provided in Fig. 7.9.

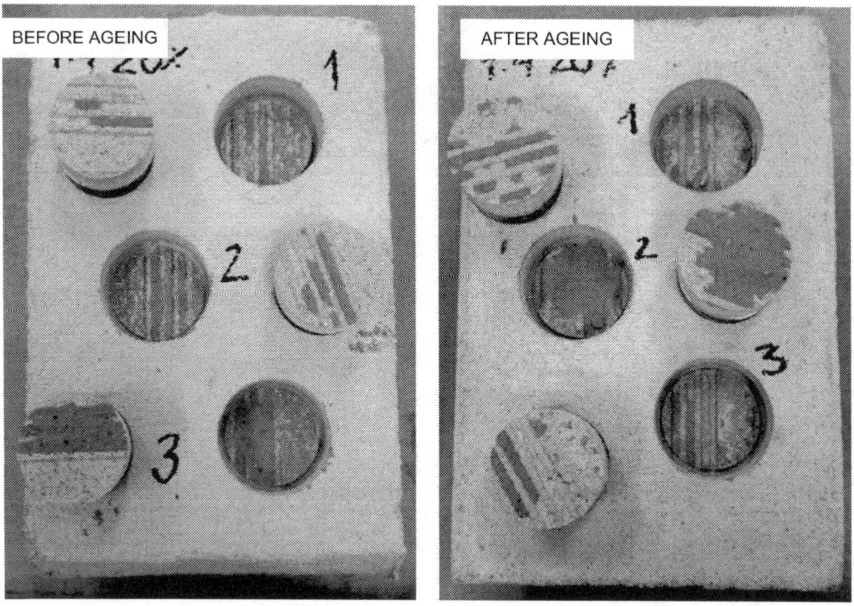

Fig. 7.6 Adherence strength fracture pattern of a mortar with sanitary ware powder waste (20%)—before and after ageing [7].

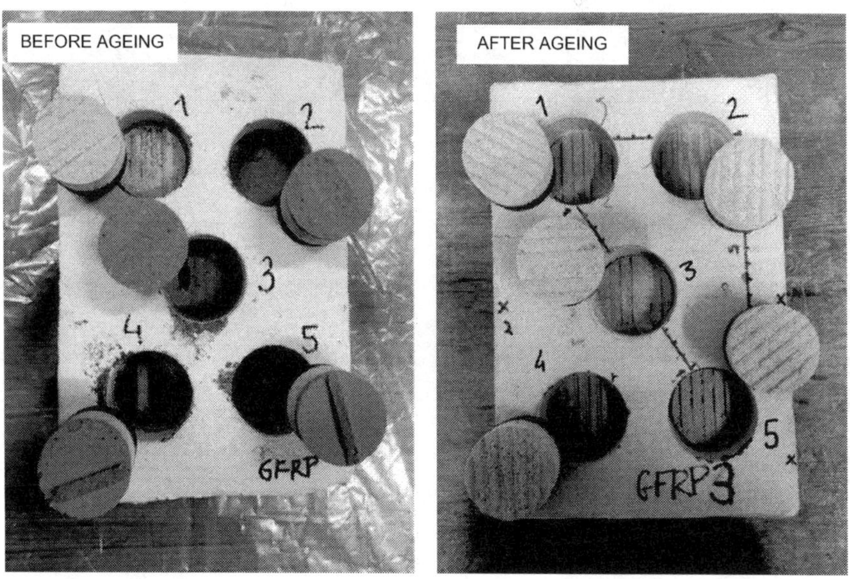

Fig. 7.7 Adherence strength fracture pattern of a mortar with GFRP powder waste (50%)—before and after ageing [218].

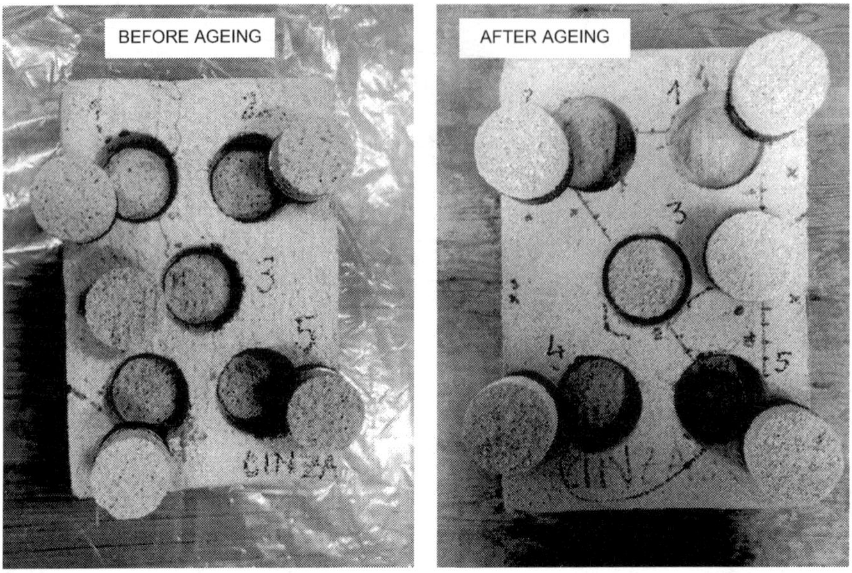

Fig. 7.8 Adherence strength fracture pattern of a mortar with forest biomass ashes (15%)—before and after ageing [218].

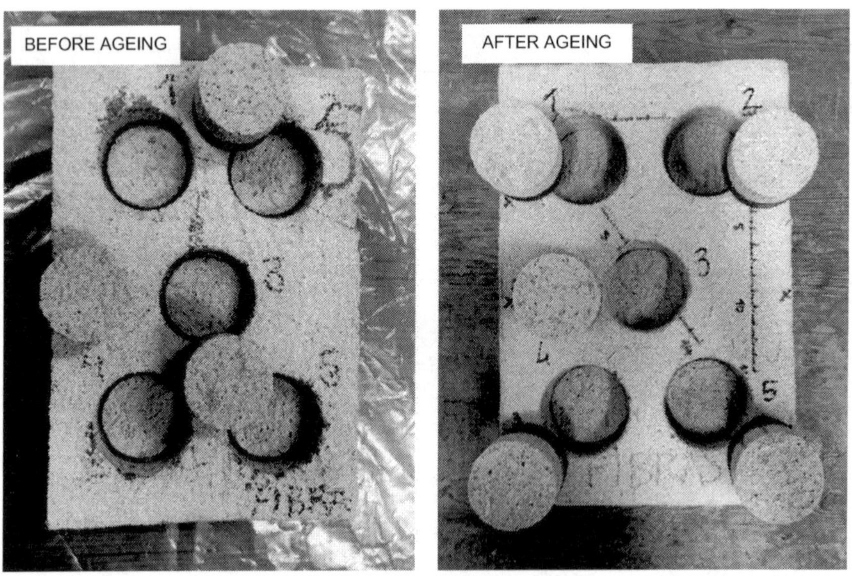

Fig. 7.9 Adherence strength fracture pattern of a mortar with textile fibers (5%)—before and after ageing [218].

As mentioned in Section 5.1.3, the adherence strength of the mortar should meet the minimum requirement to be applied as a render, which, according to the relevant National Report from LNEC [247], is defined as higher than 0.30 MPa or cohesive fracture. Considering this requirement, only the mortars with PET (flakes or pellets) incorporation did not accomplish these minimums. However, the fracture pattern was not indicated by the authors [137]; Thus, if the mortars presented a cohesive fracture, they could also be applied as render mortars according to this report.

Although all the mortars presented suitable values for render mortars application, the modification of strength with ageing gives an overall indication of the mortars' durability. In Fig. 7.10, some examples of the adherence strength obtained before and after ageing are represented.

Some mortars, when submitted to an artificial ageing, increment their adherence strength and others reduce it. Red ceramics incorporated as an aggregate [221] or as a filler [49] increment significantly the mortars' adherence strength after ageing, while GFRP, for example, considerably reduces their adherence strength value. There are other wastes that maintain their adherence practically unchanged after ageing, such as mortars' reinforced with textile fibers [218].

During render mortars' lifetime, they are submitted to several environmental conditions. In artificial ageing, extreme conditions of humidity-freeze and heat-freeze were simulated. During that process, the mortars can suffer different reactions. For instance, in humidity-freeze cycles, mortars are submitted to 8 hours of a humid environment in which water is falling into the specimens continuously. This water availability can be responsible for the late hydration of cement constituents that previously had not hydrated. These possible reactions improve the mortars' mechanical

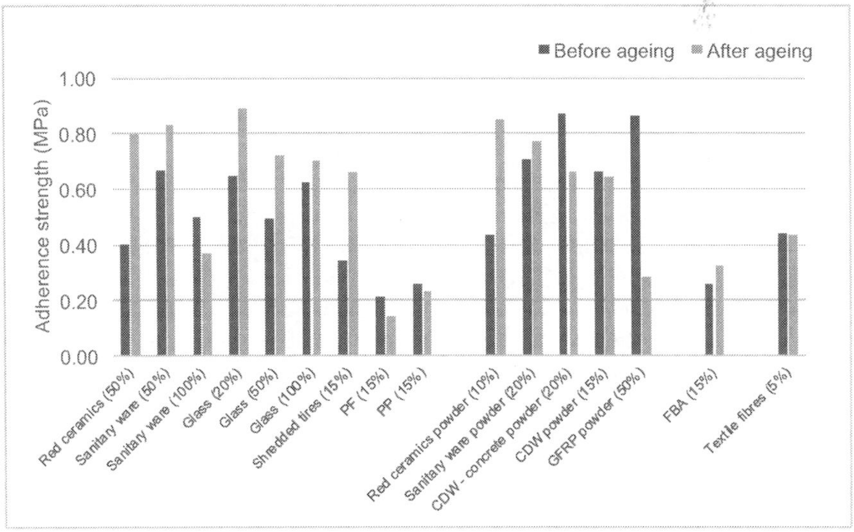

Fig. 7.10 Adherence strength before and after ageing of mortars with aggregate, filler, binder, or fiber wastes [7] [18] [49] [137] [143] [148] [154] [218] [221].

strength, which can also increase the adherence strength value. Improvements in adherence strength can be noticed in several mortars with wastes' incorporation. In aggregate wastes, this trend was noticed for red ceramics [221], sanitary ware (50%) [18], glass [143], and shredded tires [148]. In filler wastes, only for red ceramics [49] and sanitary ware [7] was an increment of strength after ageing noticed. For FBA [218], the waste that appears as an example of binder wastes, an increment of adherence after ageing was also noticed. These increments are due to the improvement of the compactness of the binder matrix of the mortars, after the late hydration of the cement and the development of the pozzolanic reaction (in the case of the ashes) in those cycles.

However, there are other modifications in mortars during artificial ageing that can be responsible for worse adherence. The thermal and humidity variations, which occur during the artificial ageing cycles, can create stresses in the interface mortar/substrate and degrade it. In the specific cases presented in Fig. 7.10, the contact area between the mortars and the substrate is low, because small specimens were analyzed: only the area of a 60 cm^2 brick face. In this low contact area, the stresses generated due to the thermal or humidity variations are probably lower than the ones that the render is subjected to when restrained by the substrate. These variations can also be responsible for the appearance of cracks in the mortars, when the tensile resistance of the mortar is exceeded to release the stresses. Internal microcracking can also occur in the specimens during humidity-freeze cycles. The water penetrates into the pores and, should the temperature turn negative, water freezes creating stresses on the surface of the pores that can produce internal microcracks. This cracking will be responsible for the degradation of mortars matrix and consequently for reducing their mechanical strengths, including adherence strength. This microcracking occurs usually in less deformable mortars, as in mortars incorporating CDW (construction and demolition waste) powder, mainly composed of concrete particles [154].

In heat-freeze cycles, an infrared radiation heats the specimens to create heat. This radiation can also be responsible for the degradation of the mortars, namely the ones with polymers incorporation. With ageing, the mortars with PET incorporation (both pellets and flakes) and the mortar with GFRP wastes see their adherence strength reduced with ageing. These reductions can be due to the modification of the polymer particles when the infrared radiation is focused on the matrix. In fact, mortars with GFRP waste, when submitted to artificial ageing, decreased their adherence strength by about 67%, from 0.86 to 0.28 MPa. In addition, a crack was noticed in the specimens after ageing, as seen in Fig. 7.29, which can indicate that simultaneous effects occur: a degradation of the matrix due to infrared radiation and probably thermal or humidity variations responsible for the mortars' cracking.

7.2.3 Impact resistance before and after ageing

Impact resistance is a mortars' characteristic that depends on the mortar's deformability and compressive strength (see Section 5.1.4). Deformability is influenced by ageing, as presented previously. In this sense, impact resistance can also be affected by ageing and can influence the render mortar's durability.

In a previous research [218], the influence of ageing on mortars' impact resistance was evaluated. Mortars containing sanitary ware waste, FBA, and textile fibers were tested. The results of each mortar are described next.

First, an aggregate waste and its influence on impact resistance after ageing is presented: sanitary ware (SW100). The waste fully replaced the natural aggregate. The results of the test are presented in Figs. 7.11 and 7.12. In the first one, the impact energy until the first crack occurs, before and after ageing, is presented and in the second failure energy is represented. The blended mortar was compared with a reference (REF) mortar composed of sand as natural aggregate.

The impact energy necessary to produce the first crack was similar (before ageing) between the REF and SW100 mortars, about 10.78 J. After the ageing process, the REF mortar needed less energy to achieve the first crack. The SW100 mortar, on the contrary, increased the impact energy after ageing. Concerning the dent produced by the mass fall, before and after ageing, the SW100 mortar presented dents with higher diameters, lower depth, and lower width. These results indicate that the SW100 mortar is more deformable than the REF mortar. In addition, it was also concluded that the ageing process does not negatively influence this property of the mortar. In Figs. 7.13 and 7.14, it is possible to observe the first crack in both mortars before and after ageing.

After the first crack occurred, the results were evaluated, and the impact resistance test continued until failure. Before ageing, it was noticed that the SW100 mortar needed less energy until failure, in comparison with the REF mortar. After ageing, the opposite occurred. The SW100 mortar needed more energy until failure than the REF mortar, about 15.68 J versus 15.19 J. In addition, the REF mortar presented a single and larger width crack when failure occurred, which is characteristic of a brittle failure (Figs. 7.15 and 7.16). The SW100 mortar presented several thin well-distributed cracks over the specimen, which indicate a more ductile behavior than the REF mortar's.

The second mortar to be presented is one with incorporation of an FBA that replaced cement at 15% (by volume). In the figures, this blended mortar is designated as 15FBA. Fig. 7.17 presents the values of impact energy before and after ageing to achieve the first crack and in failure. The characteristics of dents obtained in the first crack are summarized in Fig. 7.18. The results and characteristics are in comparison with a REF mortar (without wastes' incorporation), designated as REF.

The impact energy necessary for the first crack to occur (Figs. 7.13 and 7.19) was modified in both mortars by the ageing process. The REF mortar presented a 5% lower impact energy after ageing (10.8 J vs 10.3 J), which is not a significant reduction. The FBA mortar, on the other hand, presented a higher impact energy after ageing (11.8 J vs 12.7 J), increased by about 8%. The impact energy differences before and after ageing represented an improvement of deformability for the FBA mortar and a decrease for the REF mortar, although the differences were not significant. The crack width decreased after ageing for both mortars under analysis.

The diameter and depth of the dent of both mortars were also evaluated after ageing. A slight decrease in diameter was noticed after ageing: in the 15FBA mortar, the reduction was 7% and in the REF mortar it was only 2%. A decrease of dent depth in

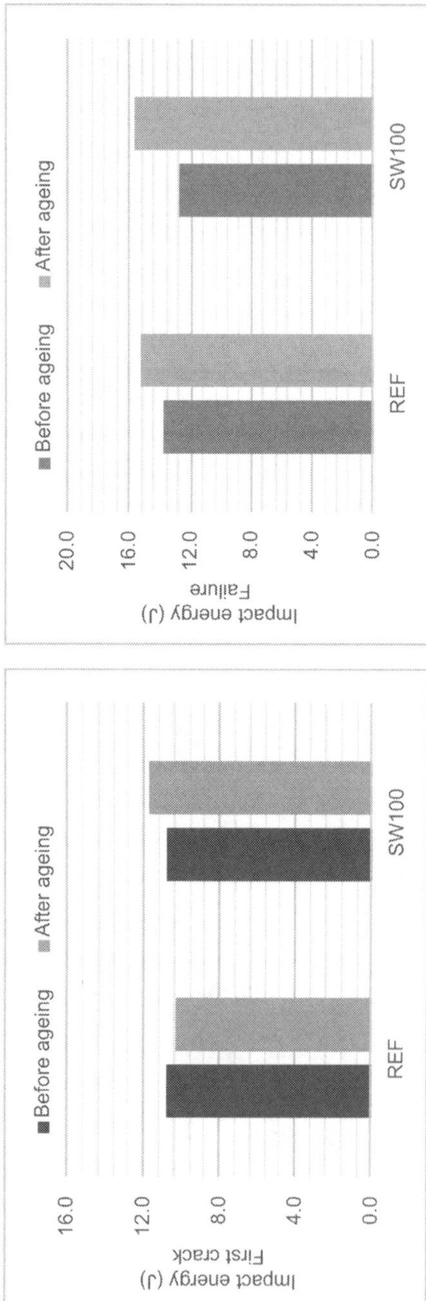

Fig. 7.11 Impact energy before and after ageing of SW100 mortar [218].

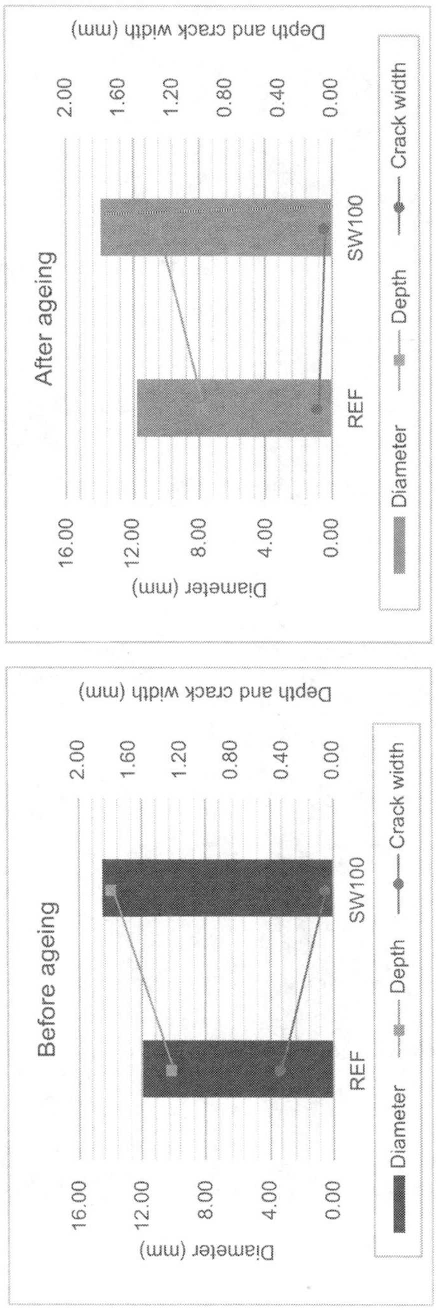

Fig. 7.12 Impact dent characteristics and cracking width before and after ageing of SW100 mortar [218].

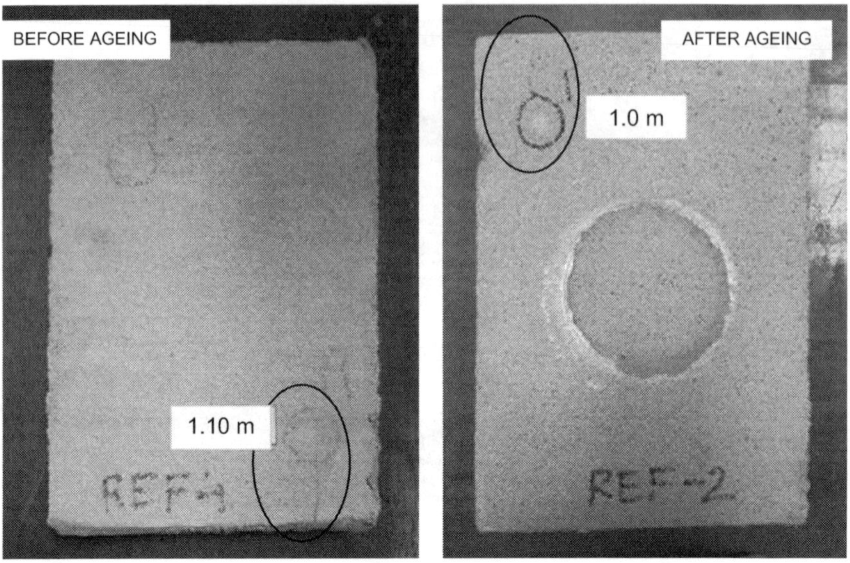

Fig. 7.13 Resistance to impact (first crack) before and after ageing of REF mortar [218].

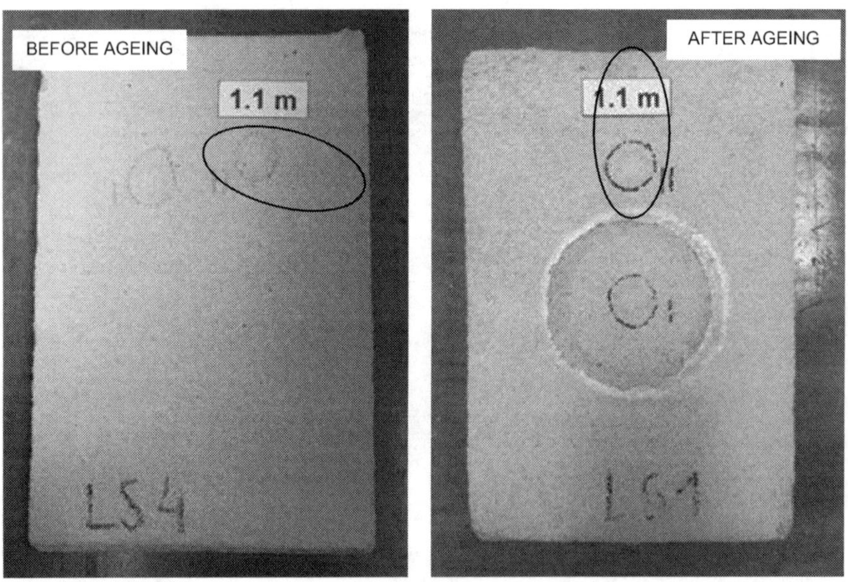

Fig. 7.14 Resistance to impact (first crack) before and after ageing of SW100 mortar [218].

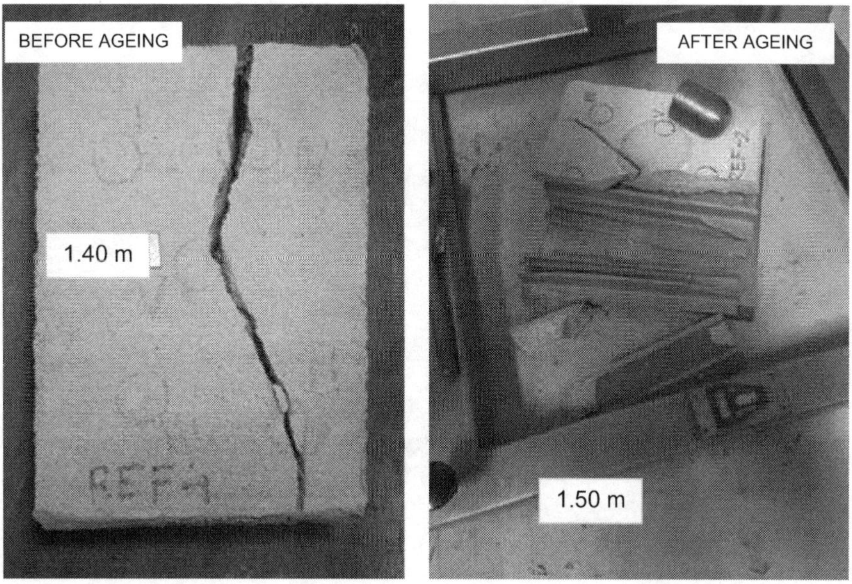

Fig. 7.15 Resistance to impact (failure) before and after ageing of REF mortar [218].

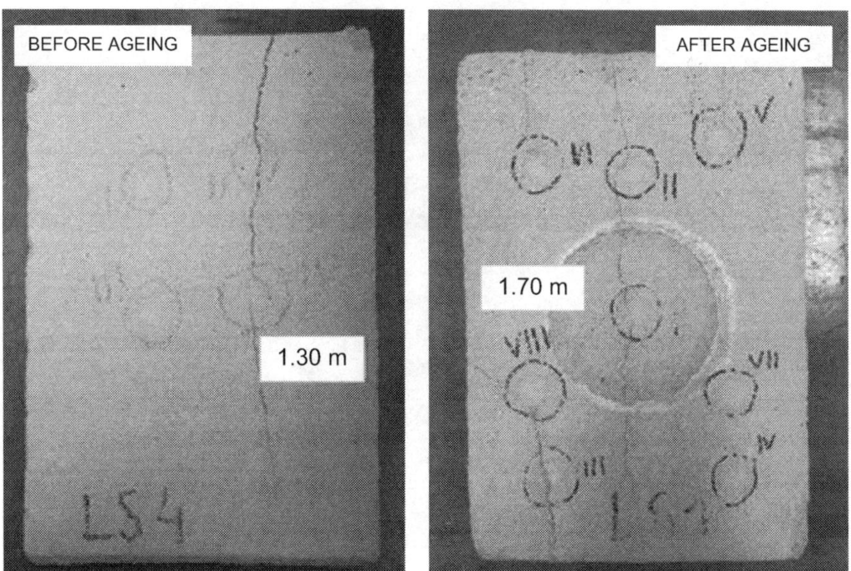

Fig. 7.16 Resistance to impact (failure) before and after ageing of SW100 mortar [218].

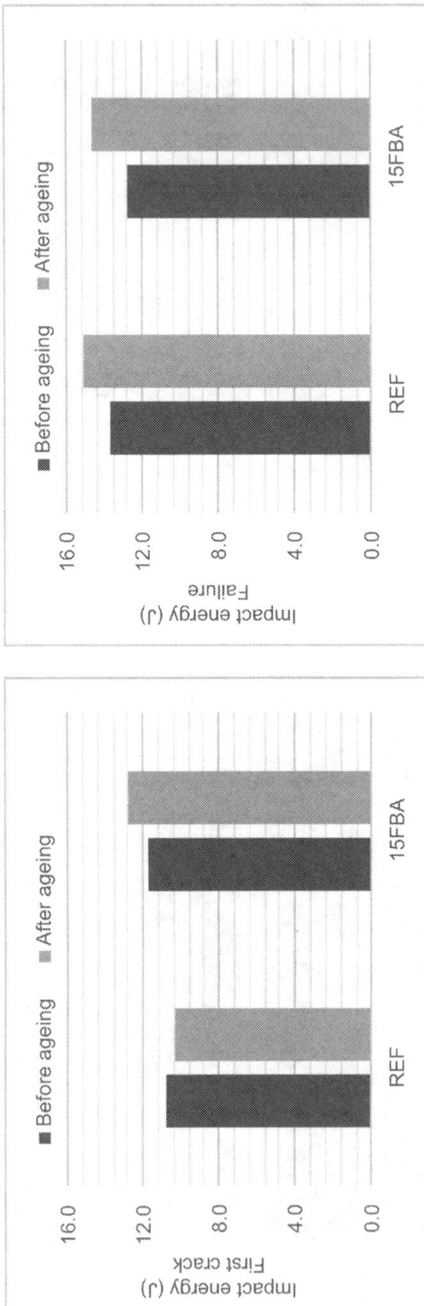

Fig. 7.17 Impact energy before and after ageing of the FBA waste mortar [218].

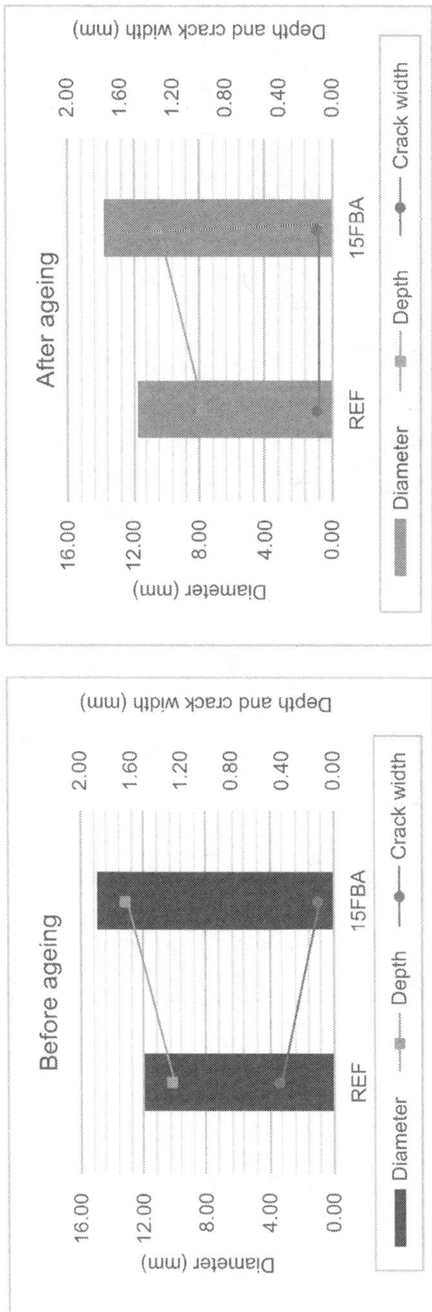

Fig. 7.18 Impact dent characteristics and cracking width before and after ageing of the FBA waste mortar [218].

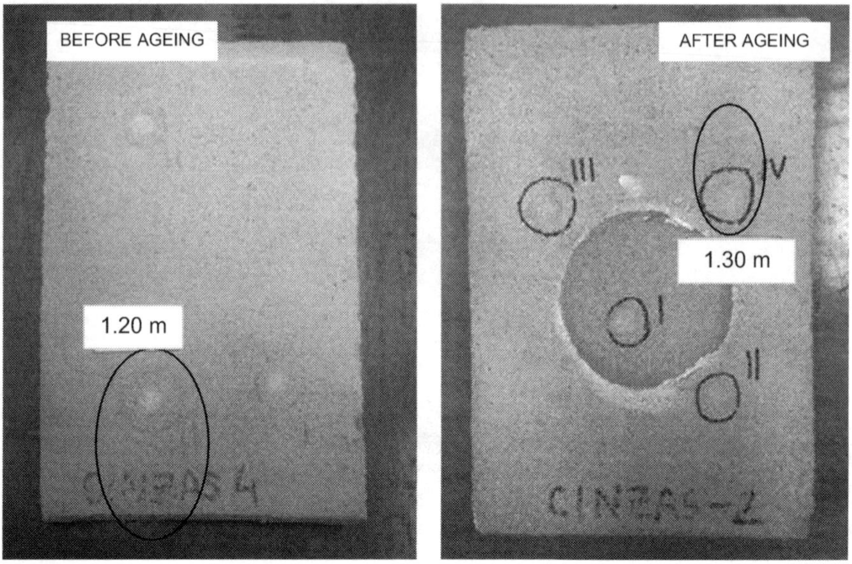

Fig. 7.19 Resistance to impact (first crack) before and after ageing of 15FBA mortar [218].

both mortars was also noticed: the 15FBA mortar decreased the dent by 20% (1.63 mm vs 1.30 mm) and REF's decreased it by about 22% (1.25 mm vs 0.97 mm). A reduction in diameter and depth of the dent represents a decrease in deformability, which is in accordance with the critical height (in the first crack) of the REF mortar but disagrees with what was noticed in the FBA mortar.

The impact energy, when fracture occurred, was higher after ageing for both mortars (Fig. 7.15 and Fig. 7.20). The REF mortar presented an increase of energy of about 11% (13.7 J vs 15.2 J) and the 15FBA mortar an increase of 15% (12.7 J vs 14.7 J). The increment of energy indicates higher mechanical strength after the ageing process, which can be due to late hydration of the cement components, as indicated previously in the adherence strength analysis. Comparing the failure before and after ageing, probably due to the improvement of mechanical strengths and the reduction of deformability, the failure is more brittle after ageing (Figs. 7.15 and 7.20), and more evidently so for the REF mortar.

The influence of ageing in a mortar reinforced by textile fibers is also illustrated as an example. In Figs. 7.21 and 7.22, the values obtained in impact resistance test by Farinha [218] are illustrated.

It was noticed that the incorporation of textile fibers improved the impact resistance, when compared with a current mortar (REF mortar), until the first crack arises, due to the higher deformability of these modified mortars. As far as the impact resistance until the first crack is concerned, a reduction after ageing was found, for both mortars, which is in accordance with the decrease of deformability found in the ultrasound pulse velocity test. In fact, the reduction of deformability already addressed has a negative influence on the impact energy; however, even after ageing, the mortar

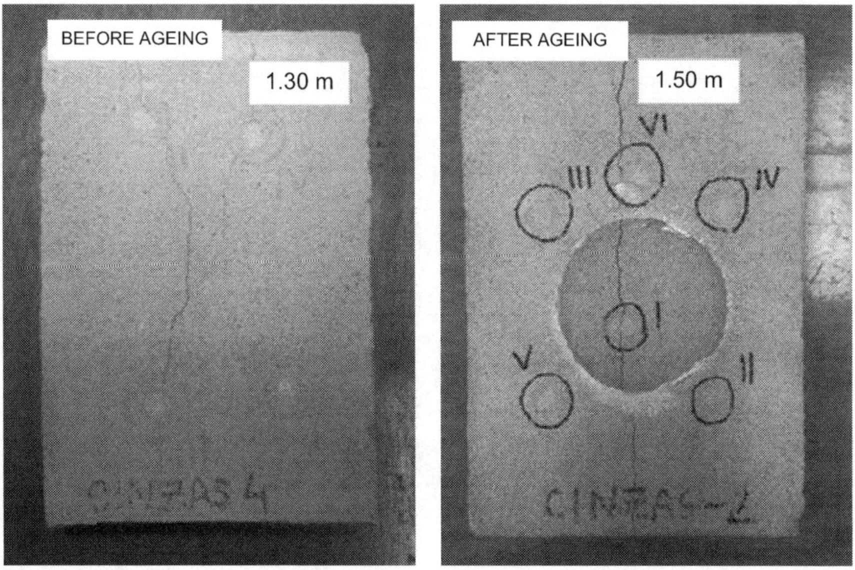

Fig. 7.20 Resistance to impact (failure) before and after ageing of 15FBA mortar [218].

with textiles presented a higher resistance than REF's. Figs. 7.13 and 7.23 illustrate the first cracks before and after ageing of the REF and TF1.5 mortars, respectively. The falling mass created a dent in the specimens, the diameter and depth of which were also measured, as well as the crack width of the first crack. Before ageing, an increase of the diameter and depth in the textile specimen was noticed, as well as a lower crack width. After ageing, the scenario was different: the diameter and depth of the TF1.5 mortar was slightly reduced in comparison with the REF mortar and the cracking width was similar in all mortars.

Concerning the impact energy in failure, it was noticed that the REF mortar increased its energy and the TF1.5 mortar maintained the value obtained before ageing. In terms of adherence strength, an increase of strength with ageing was noticed in the REF mortar and a similar value with ageing in the TF1.5 mortar, which is in accordance with the impact resistance test. Ageing improved the REF mortar strength and did not alter the strength of the TF1.5 mortar. Figs. 7.15 and 7.24 illustrate the failure cracks before and after ageing in the REF and TF1.5 mortars, respectively. In the figures, it is possible to identify a brittle failure of the REF mortar. Before ageing, the mortar presented a single crack with a large width that crossed the entire specimen; after ageing, the specimen was divided into several pieces. Both failures are characteristics of a brittle material. The mortar reinforced with fibers presented a totally different failure pattern. Before ageing, the TF1.5 mortar presented distributed cracks with low widths. It was also noticed that the ceramic substrate, where the mortar was applied, presented a more brittle behavior than the mortar's. After ageing, a similar behavior of the TF1.5 mortar was noticed: distributed cracks with small widths. These failure patterns are characteristics of a ductile material.

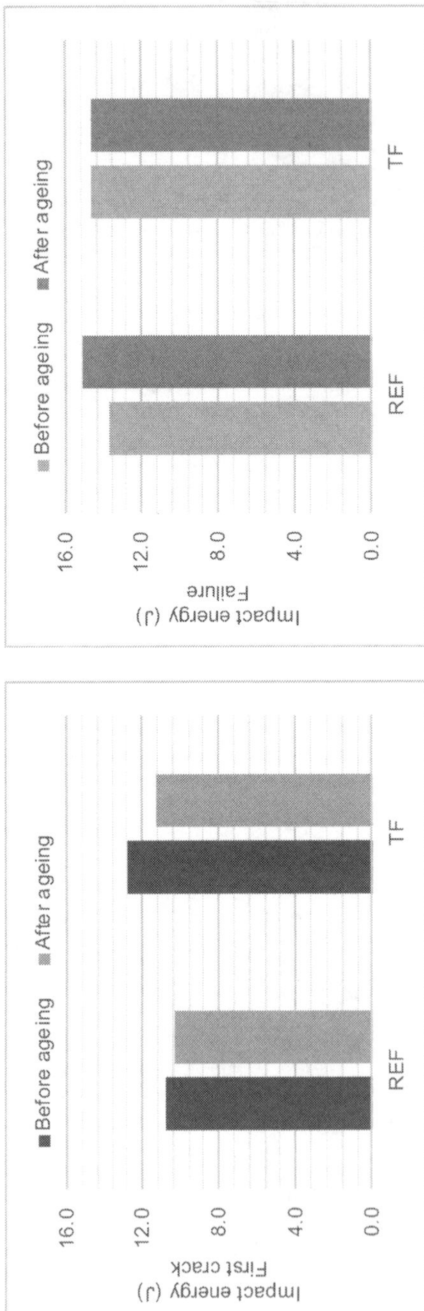

Fig. 7.21 Impact energy before and after ageing—TF 1.5 mortar [218].

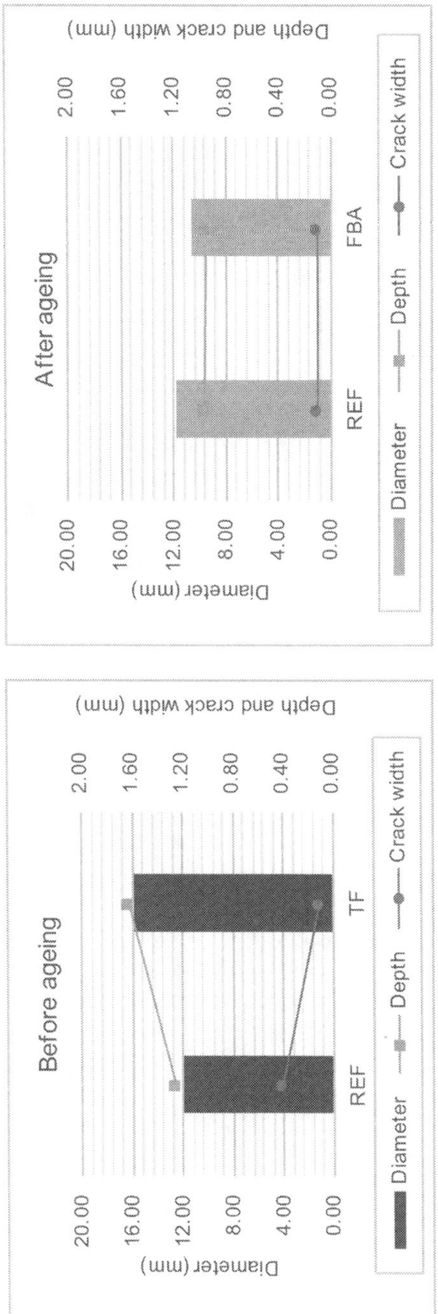

Fig. 7.22 Impact dent characteristics and cracking width before and after ageing—TF 1.5 mortar [218].

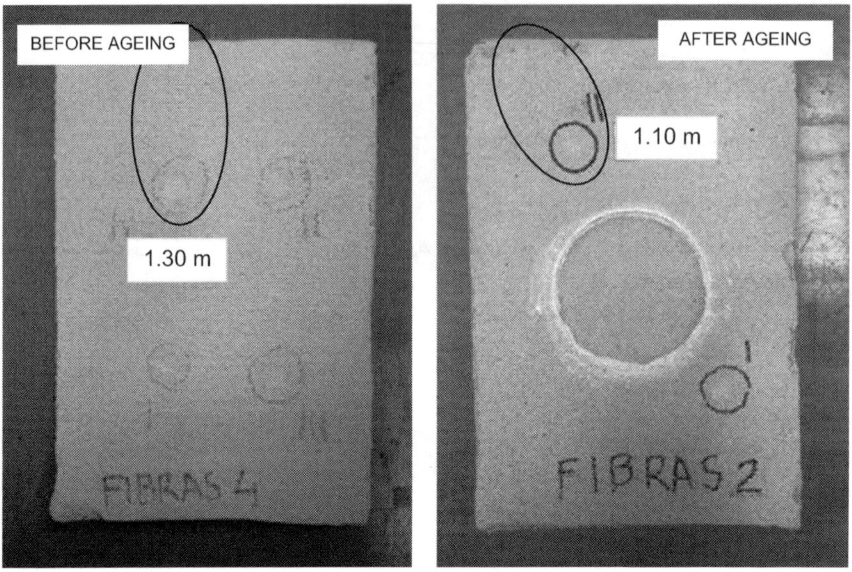

Fig. 7.23 Resistance to impact (first crack) before and after ageing—TF 1.5 mortar [218].

The mortar reinforced with fibers did not significantly change its properties with ageing. It slightly decreased its deformability, which slightly reduced the impact resistance until the first crack. However, the other properties analyzed were not modified: similar adherence strength before and after ageing, water absorption under pressure, and failure pattern and resistance to impact were noticed. It can be concluded

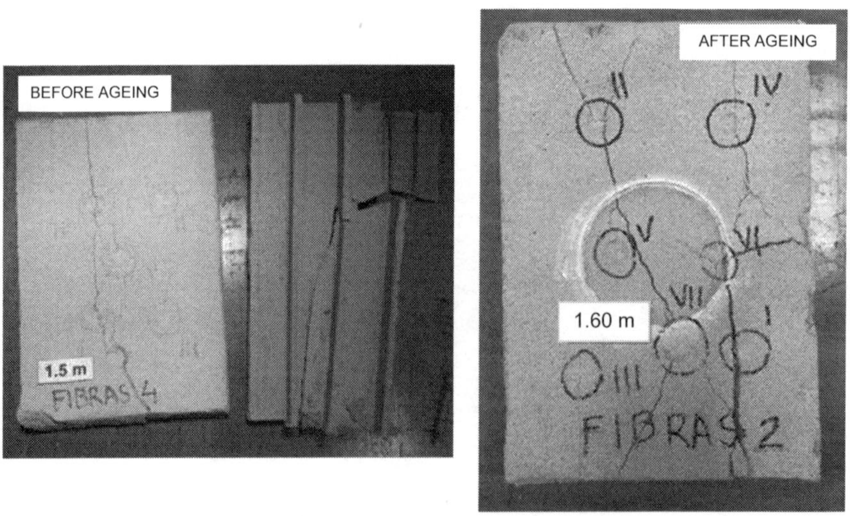

Fig. 7.24 Resistance to impact (failure) before and after ageing—TF 1.5 mortar [218].

that the incorporation of fibers maintained the durability of the mortars during the ageing process, unlike for the REF mortar.

7.2.4 Permeability to water before and after ageing

Mortars' permeability is a characteristic that depends on the mortars' volume of pores and their diameter. Under pressure, permeability to water is also influenced by the presence of cracks. A modification of the mortars' microstructure influences the mortars' permeability. In this sense, permeability is a characteristic that can be modified over time.

As discussed throughout this book, a render mortar should have low permeability to liquid water and, whenever the environmental conditions allow it, should release that same amount of water avoiding water remaining for a long period of time in the interior of its pores.

Next, examples of mortars with wastes' incorporation and their influence on permeability to water under pressure after artificial ageing are presented. Similarly to adherence strength, water permeability under pressure tests can give information about the mortars' durability. By comparing the results before and after ageing, it is possible to conclude about the performance of the mortars over time. In the examples presented next, mortars were submitted to extreme conditions of humidity-freeze and heat-freeze.

In Table 7.7, the permeability of water under pressure of mortars with wastes' incorporation is quantified. The relation between the values obtained by the blended mortars and their correspondent REF mortar is illustrated in Fig. 7.25.

Permeability to water under pressure depends on the type of wastes' incorporation. Mortar with aggregate wastes presents greater permeability to water than the

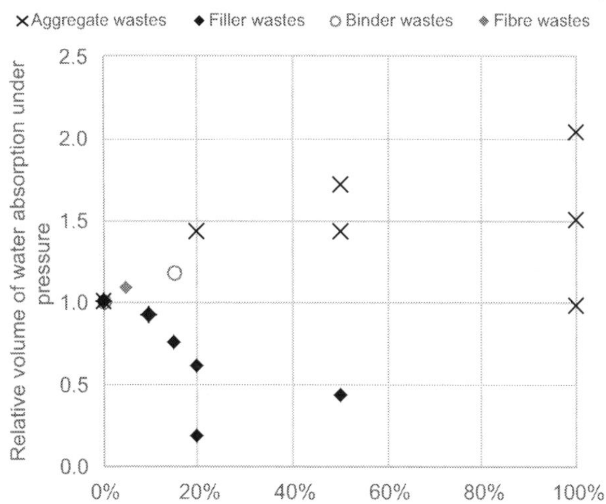

Fig. 7.25 Relative permeability to water under pressure of mortars with aggregate, filler, binder, or fiber wastes after ageing [7] [18] [49] [143] [154] [155] [218] [221].

corresponding REF mortar. Sanitary ware was the waste that replaced sand at 100% [218] producing a mortar with permeability closer to that of the REF mortar. Glass waste used as the only aggregate [143] incremented this permeability by more than 100%, in comparison with the REF mortar, achieving a value of 3 ml/cm^2 while the REF mortar absorbed 1.5 ml/cm^2. On the other hand, when the wastes are incorporated as fillers, the reduction of permeability is noticeable. Sanitary ware powder, when incorporated by 20% (in volume), reduces the mortars' permeability from 4.33 ml/cm^2 to 0.77 ml/cm^2, corresponding to a reduction of more than 80%. The binder waste, presented as an example, incremented by 17% this permeability, after ageing. Textile fibers waste was also responsible for an increment of the permeability; however, this increment was as low as about 7%. It is necessary to mention that the mortars presented as examples were rather poor in cement, with volumetric ratios of 1:4, as is usually the case with render mortars, and with no admixtures incorporated, such as water repellent. These low cement contents and the inexistence of admixtures are responsible for values of permeability relatively high; however, it allows evaluating the effective performance of the wastes' particles.

In Fig. 7.26, the results of permeability to water under pressure before and after ageing for some mortars with wastes' incorporation are illustrated. In general, mortars with waste incorporation increment their permeability to water under pressure after ageing. Only the incorporation of sanitary ware as an aggregate and replacing sand at 100% resulted in a mortar less permeable than it was before ageing. As mentioned in the adherence strength analysis after ageing (Section 7.3.2.), there are some modifications on the mortars after ageing, which include cracking, internal microcracking, and late hydration of the cement components. This late reaction that can occur is

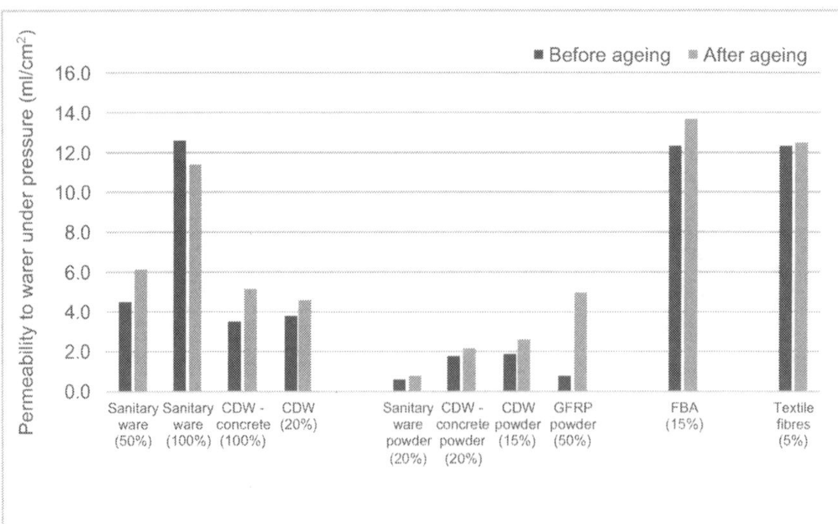

Fig. 7.26 Permeability to water under pressure before and after ageing of mortars with aggregate, filler, binder, or fiber wastes [7] [18] [49] [143] [154] [155] [218] [221].

responsible for an increment in mortars' compactness and consequently can reduce mortars' permeability. The remaining modifications in mortars as cracking due to thermal or humidity variations, to movements of the substrate, to strokes or others as internal micro-cracking as the one that occurs due to the formation of ice in the interior of the pores are all responsible for a reduction of the mortars' compactness. This modification of mortars' microstructure influences the mortars' permeability, namely increasing it, which can be noticed in the graph presented in Fig. 7.26. The increments vary in general between 1% and 46%, depending on the mortars. Textile fibers incremented by only 1% the permeability after ageing [218]. The CDW mainly composed of concrete incremented the permeability by about 46% [155] and the FBA by about 11% [218]. The mortar that contained GFRP waste significantly incremented their water permeability after ageing, by about 588% [218]. A degradation of the waste in artificial ageing, due to the infrared radiation and probably due to the humidity-freeze cycles was the main reason for a general microcracking, both internal and on the surface (see next section). From the examples presented, in fact GFRP was the one that presented the lowest durability.

7.3 Esthetical appearance

Esthetical appearance is an aspect that concerns inhabitants in general [77]. A construction should present a friendly environment to the users. A worse esthetical appearance can make them feel uncomfortable in their dwellings or offices and therefore it should be avoided. Several anomalies can occur that affect the renders' esthetical appearance. The modification of color of render mortars, the appearance of stains (light or dark), cracking, or bio-colonization are examples of esthetical modifications that can affect the buildings global appearance.

The esthetical appearance of renders due to ageing, cracking, or bio-colonization is discussed in the following sections, as well as the influence of wastes' incorporation on those phenomena.

7.3.1 General esthetical appearance after ageing

Esthetical appearance of render mortars with wastes' incorporation was evaluated after artificial ageing by several authors. In Table 7.8, the appearance of these mortars, namely in terms of microcracking or leaching of the upper coating layer, is summarized. The mortars, enumerated as examples, were applied over a ceramic substrate and were submitted to artificial ageing. This ageing procedure consisted of eight heating-freezing plus eight wetting-freezing cycles. After that, the upper layer of the mortars was analyzed, in comparison with the appearance before ageing.

In general, no cracks were visible to the naked eye in the surface of the specimens (Figs. 7.27– 7.31). Only in the mortar with GFRP waste filler was a crack detected [218]. This microcracking extended over the entire length of the specimen and it is represented in red in Fig. 7.29. The resin present in the glass fiber polymer probably deteriorated during the heat-freeze cycles, leading the mortar to crack. The increment

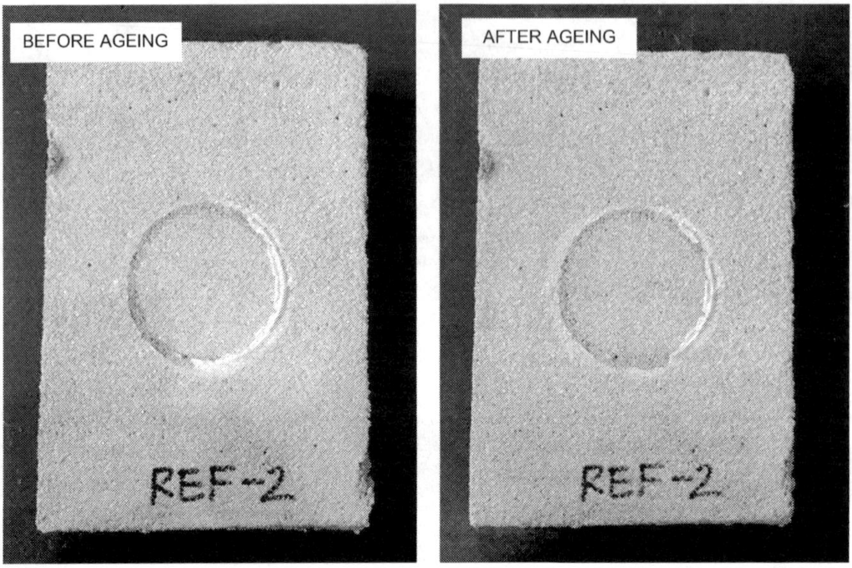

Fig. 7.27 Esthetical appearance of a reference mortar (before and after artificial ageing). Anomaly detected: leaching of the upper coating layer [218].

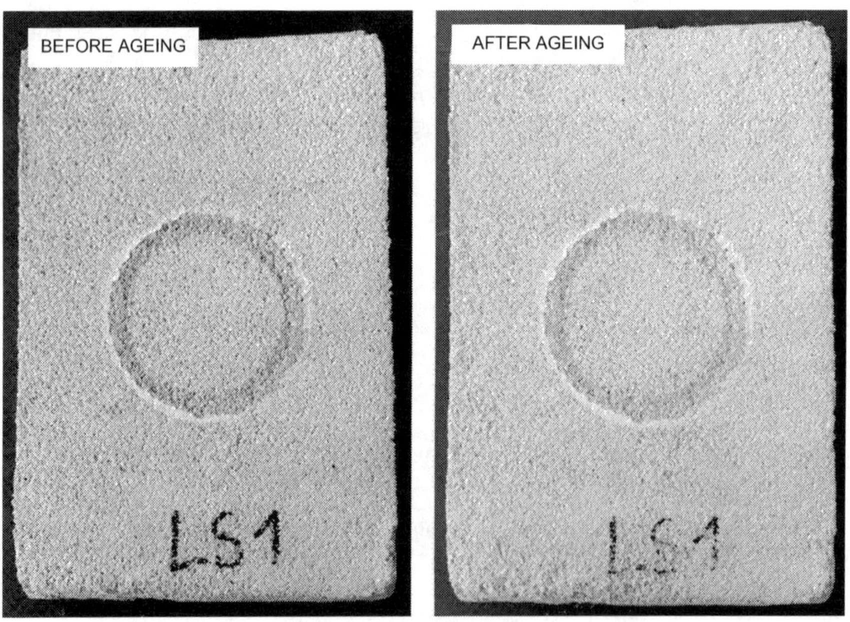

Fig. 7.28 Esthetical appearance of a mortar with sanitary ware waste incorporation as an aggregate (before and after artificial ageing). Anomaly detected: leaching of the upper coating layer [218].

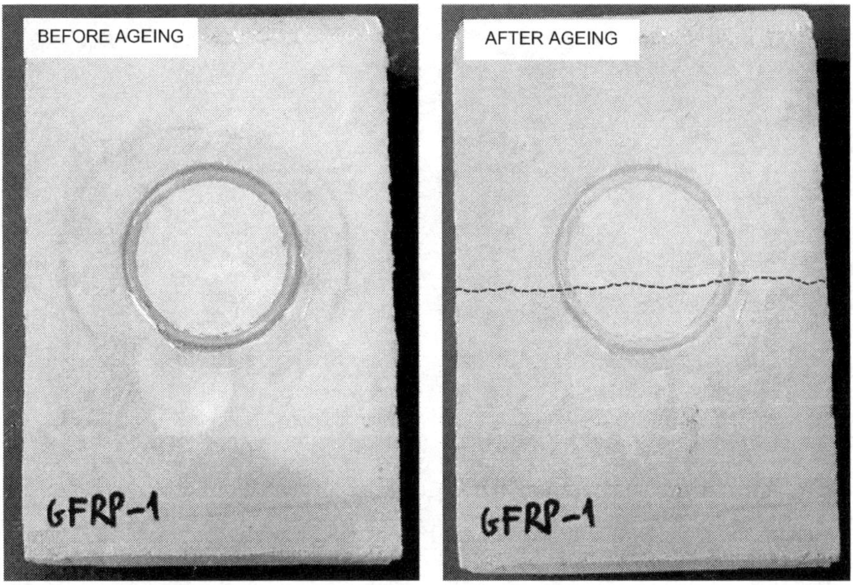

Fig. 7.29 Esthetical appearance of a mortar with GFRP powder incorporation (before and after artificial ageing). Anomalies detected: leaching of the upper coating layer and mortars' cracking [218].

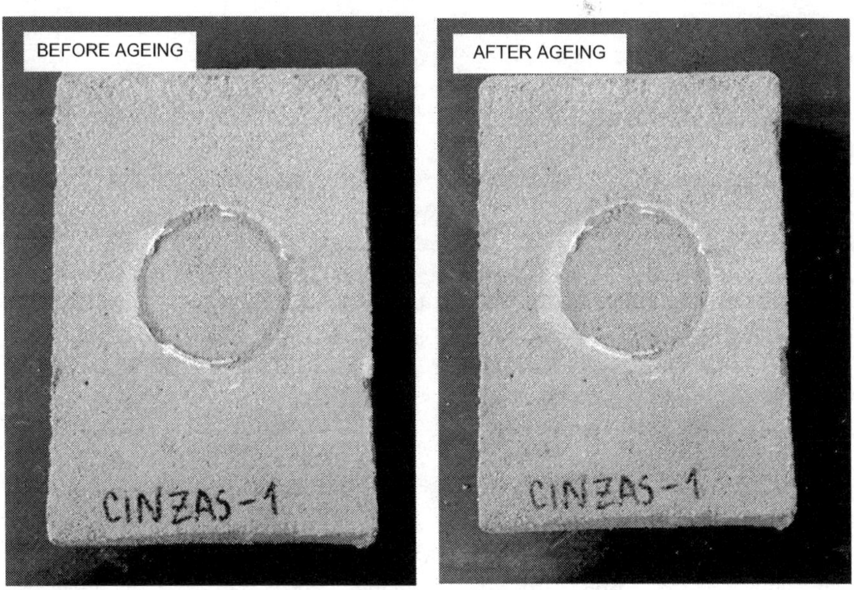

Fig. 7.30 Esthetical appearance of a mortar with forest biomass ashes incorporation (before and after artificial ageing). Anomaly detected: leaching of the upper coating layer [218].

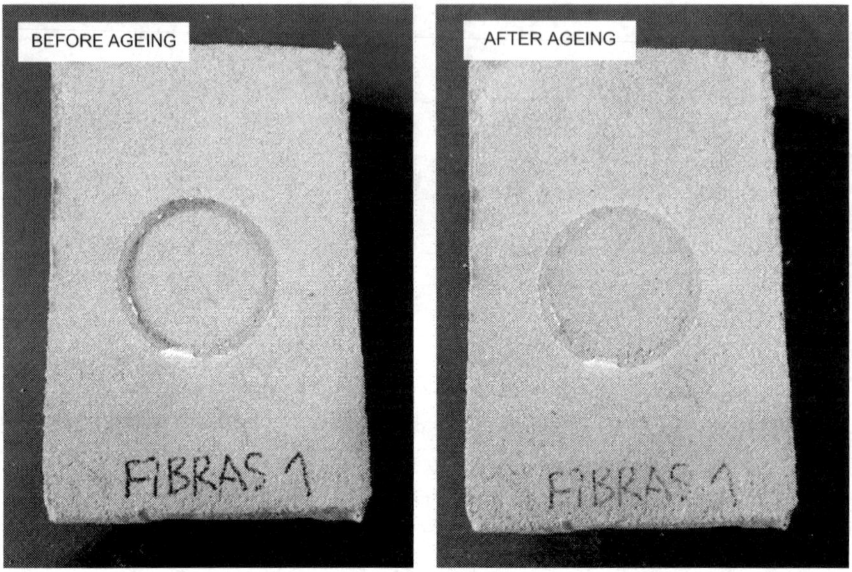

Fig. 7.31 Esthetical appearance of a mortar reinforced with textile fibers incorporation (before and after artificial ageing) [218].

of ME, during the ageing process, could also have contributed to microcracking susceptibility. Microcracking affected the esthetical appearance of the render, but it was not the only characteristic; the permeability of the render to water was also affected and can lead to other anomalies as it will be presented later in this chapter.

Although most of the specimens did not present visible microcracks, the leaching of the upper coating layer was noticed by some authors [7] [18] [137] [154] [218]. This leaching occurs during the wetting-freezing cycles. The continuous incidence of liquid water over the specimens leads to dissolution of some soluble compounds such as calcium hydroxide. This leaching often produces whitish stains and is responsible for a slight reduction of the upper layer thickness; indeed, the effect can easily be noticed. For example, in Fig. 7.31 after ageing the textile fibers are more visible than before ageing. This effect is noticed in the blended mortars but also in the REF mortar, meaning that this effect is to be expected and the incorporation of wastes does not increment it.

7.3.2 Cracking

Cracks appearance, as described in Section 2.3.5, can occur due to numerous reasons. Faulty design, inadequate application, thermal variations, humidity variations, ice formation, salt formation are some examples that can cause render to crack.

Regardless of the reason that leads to cracking, its occurrence is an anomaly that can affect the construction's global esthetical appearance. Its impact becomes higher with the cracking width and extension throughout the surface. Depending on the cracks width, they can be classified as microcracks, medium cracks, or large cracks [77]. Following the referred classification, a microcrack is a linear opening with a width lower than 0.2 mm [77], medium cracks present a width between 0.2 and 2 mm and large cracks are characterized by widths over 2 mm [77]. Microcracks are almost unnoticeable to the naked eye, and only an observer at a short distance from the wall can notice their occurrence. Medium and large cracks, on the other hand, can already be observed from a considerable distance. The texture of the render also makes cracks more or less noticeable and therefore can influence the esthetical global appearance of the renders. In a more textured finishing coating, the cracks are less visible than in a smooth finishing coating.

Cracking, whenever it occurs, should be repaired to not affect the render's durability [89]. This anomaly can affect the water tightness and consequently its protection ability [77], as it creates an open path to liquid water's ingress. This water availability can also be responsible for other associated anomalies, such as dissolution of salts, susceptibility to the growth of plants, or biocolonization.

In Section 5.2, it was mentioned that some wastes' types, such as fillers, are more susceptible to mortar cracking. These types of waste often increment the ME of the mortars, as well as their mechanical strength, decreasing the mortars' deformability. This reduction can be responsible for the mortars' cracking and consequently can influence the esthetical appearance of the render. On the contrary, aggregate, binder, or fiber wastes can reduce mortars' cracking due to the lower ME of the mortars and, in the case of fibers, due to a bridging effect. Fiber wastes are the ones that contribute the most to cracking inhibition. This type of wastes minimizes the appearance of cracks and whenever these cracks occur, this type of wastes reduce their widths and propagation.

7.3.3 Biocolonization

Biocolonization affects the general esthetical appearance of buildings. Molds proliferation in coatings is a common phenomenon that affects their general appearance. These microorganisms are filamentous fungi that develop mainly in moist environments rich in organic matter. The presence of this matter combined with a temperature of about 20–25 °C, with high relative humidity, allows the proliferation and growth of molds [227].

This proliferation is characterized by a modification of renders color and texture. Coloration depends on the type of mold. In Figs. 7.32 and 7.33, two examples of biocolonization are represented. As noticed, the appearance of the walls was clearly affected by this anomaly. This problem affects the global appearance of constructions and therefore should be addressed. In some cases, biocolonization is also responsible for the loss of cohesion of the mortar and in general it contributes to retain water.

Biocolonization can be evaluated in an outdoor application. In this type of application, the orientation of the wall chosen should have in consideration the solar radiation that does not have direct incidence to achieve the proper conditions to molds development. However, the appearance of this type of microorganisms can take some time to take place. Alternatively, the biocolonization susceptibility can be evaluated in

Fig. 7.32 Example of a render mortar contaminated by biocolonization (example 1)

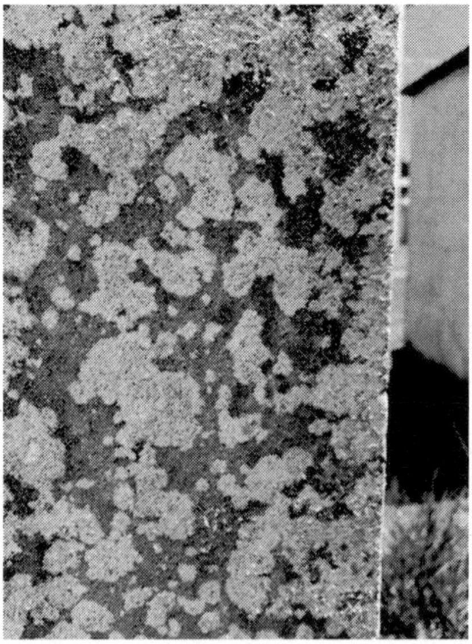

Fig. 7.33 Example of a render mortar proliferated by biocolonization (example 2)

laboratory [227]–[231]. This evaluation consists in placing the sample in a recipient in a propitious environment for molds development (that can be composed by sugar and jelly). A suspension of molds can be applied over the specimens and for a period of 4 weeks the molds development is monitored [227]–[229]. The degree of development of molds on the surface of the specimens can be quantified weekly throughout these 4 weeks of exposure, as suggested in American Standard ASTM D5590 [91].

The incorporation of organic or partially organic wastes in the mortars can be responsible for an increment of biocolonization susceptibility, as the molds feed on this type of matter. Next, some examples of wastes with organic matter and their influence on the biocolonization development are described. For all the examples, two different scenarios were investigated. In the first one, the specimens were sterilized and the inoculum of *Aspergillus niger* and *Penicillium Funiculosum* mold was suspended over the specimens. In the second scenario, the specimens were not sterilized, and no inoculum was suspended. A blank test with a filter (Whatman No. 1) was also analyzed aiming at checking the viability of the test. The sets were maintained in a chamber at a temperature of 22 ± 1 °C and a humidity of 70 ± 5% for 4 weeks. Every week, the degree of mold development on the surface of the specimens was analyzed and quantified according to the following methodology previously defined in other studies [227]–[229]: 0—no mold development; 1—less than 10% of mold development; 2—mold development between 10% and 30%; 3—mold development between 30% and 60%; 4—mold development higher than 60%. Three samples of each mortar were tested.

Polymers can originate organic wastes. Thus, an example of an organic waste that can influence biocolonization is a polymer waste. GFRP, a waste that has been presented in this book, can contain a considerable amount of organic matter. This amount depends on the proportion between the resin and the fibers. The incorporation of a GFRP waste, composed by a polyester resin (30% of volume), was researched previously [218] [229] as well as its influence on biocolonization. This waste was incorporated as a filler and the organic matter was about 12% of the total volume of the mortar. From the test, it was noticed that the development of molds in the culture environment was visible as early as the first test day (Figs. 7.34 and 7.35). In the

Fig. 7.34 Biocolonization development of a reference mortar [218] [229].

Fig. 7.35 Biocolonization development of a mortar with GFRP waste incorporation [218] [229].

containers inoculated with mixed suspension of *A. niger* and *P. funiculosum*, it was observed that both molds developed, with an apparent greater predominance of *A. niger*. In containers with natural inoculum, although *A. niger* and *P. funiculosum* were also present, the presence of an undetermined number of other molds was also noticed.

Although the importance of the *A. niger* and *P. funiculosum* fungi in aesthetic deterioration of coatings is widely recognized [229], the degree of mold development appears to be greater in the specimens in which natural inoculum growth was allowed (Fig. 7.36). Under the conditions of the present test, with the inoculum of *A. niger*

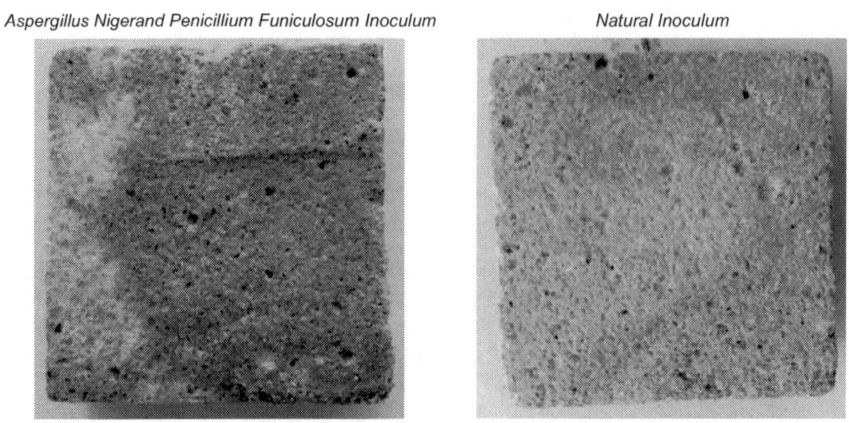

Fig. 7.36 Biocolonization development in GFRP mortar surface [218] [229].

and *P. funiculosum*, after 4 weeks of testing, only one sample of the mortar with fiber composites showed mold development (grade 2). The REF mortar did not show mold development (grade 0). On the other hand, with the *natural inoculum*, at the end of 4 weeks, mold development was observed in both mortars. The REF mortar presented values of mold contamination below 30% (grades 1–2) and mortar with GFRP waste presented the highest development, between 10% and 60% (grades 2–3) [229]. In Fig. 7.36, the molds development to the naked eye is illustrated.

The mold contamination of the composite mortar was also observed through a stereoscopic microscope, at different magnifications (Fig. 7.37).

The susceptibility to biocolonization of a mortar with FBA waste was also analyzed in a previous research [218] [229], due to the 3.2% of organic matter, which represented 0.1% of the total volume of the mortar. In Fig. 7.38, mold attacks in the blended mortar over the 4 weeks of the test are illustrated. Mold development in the culture environment was visible as early as the first test days (Fig. 7.39). In the containers inoculated with a mixed suspension of *A. niger* and *P. funiculosum*, it was observed that both molds developed, with an apparent greater predominance of *A. niger*. In containers with natural inoculum, although *A. niger* and *P. funiculosum* were also present, the presence of an undetermined number of other molds was also noticed (Fig. 7.39). In Table 7.9, the degree of development of the molds on the surface of the samples is quantified. The REF mortar in the second week of test already presented some mold development in the *natural inoculum*, until 20% of development (grades

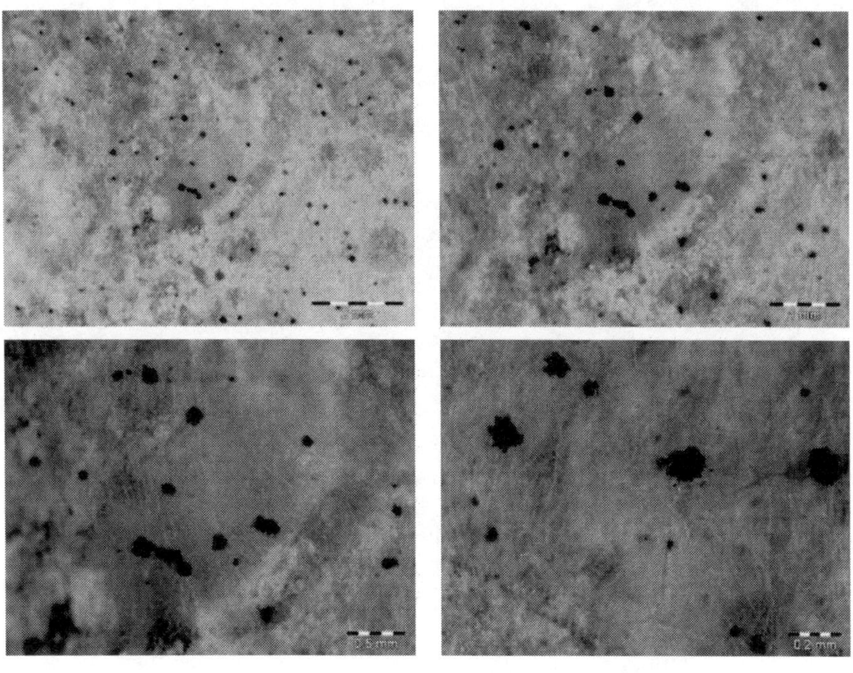

Fig. 7.37 Mold development in GFRP mortars (stereoscopic microscope) [218] [229].

Fig. 7.38 Biocolonization development in FBA mortar surface [218] [229].

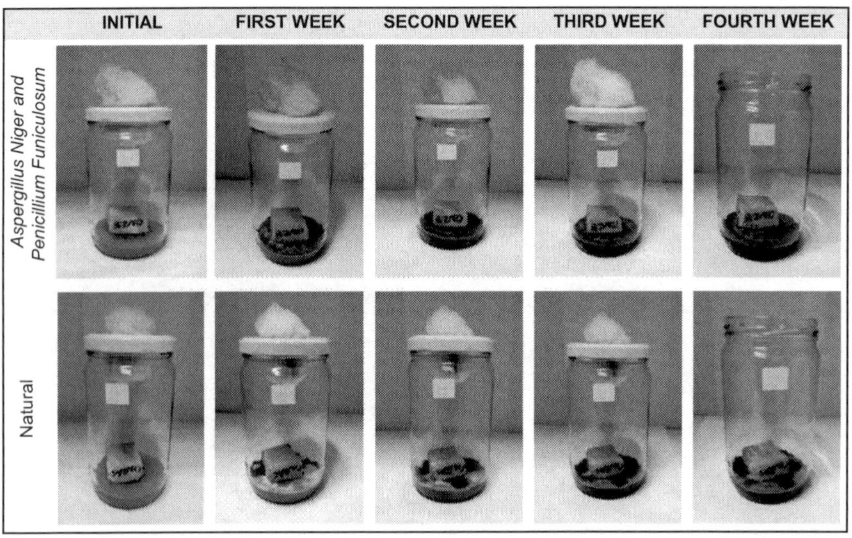

Fig. 7.39 Biocolonization development of a mortar with FBA waste incorporation [218] [229].

1–2) [229]. The FBA waste, contrary to what was expected by the authors [218] [229], even in the fourth week of test presented a degree of mold development quantified to be as less than 10% (grades 0–1), also in the *natural inoculum* (Fig. 7.38).

Similar to what was noticed for the GFRP waste, in the FBA mortar's analysis, *natural inoculum* had more relevance to the susceptibility to biocolonization than *A. niger* and *P. funiculosum* molds. Sterilization of the specimens can contribute to the results' discrepancy [218] [229], as part of the fungi favoring nutrients can be eliminated.

Table 7.9 Degree of development of molds on the surface of specimens—GFRP mortars [229].

Mortar	Inoculum	Sample	First week	Second week	Third week	Fourth week
Control	Aspergillus niger and Penicillium funiculosum	A	3	4	4	4
		B	3	3	4	4
		C	3	3	4	4
Reference	Aspergillus niger and Penicillium funiculosum	A	0	0	0	0
		B	0	0	0	0
		C	0	0	0	0
	Natural	A	0	1	1	1
		B	0	1	1	1
		C	0	2	2	2
GFRP	Aspergillus niger and Penicillium funiculosum	A	0	0	0	0
		B	0	0	2	2
		C	0	0	0	0
	Natural	A	0	0	1	3
		B	0	1	1	2
		C	0	0	1	2
Forest biomass ashes	Aspergillus niger and Penicillium funiculosum	A	0	0	0	0
		B	0	0	0	0
		C	0	0	0	0
	Natural	A	0	0	0	1
		B	0	0	0	0
		C	0	0	0	1
Textile fibers	Aspergillus niger and Penicillium funiculosum	A	0	0	0	0
		B	0	0	0	0
		C	0	0	0	0
	Natural	A	0	0	0	0
		B	0	0	0	0
		C	0	0	0	0

Note: 0—no molds growth; 1—less than 10% of molds growth on the surface of the specimen; 2—from 10% to 30% of molds on the surface of the specimen; 3—between 30% and 60% of molds on the surface of the specimen; 4—more than 60% of molds on the surface of the specimen.

Fibers are another type of wastes that can have organic content, and textile fibers are one of these. Next, an example of a textile fiber incorporated in a render mortar is illustrated [229]. The biocolonization susceptibility test lasted for 4 weeks and in Fig. 7.40 the mold development during this period is presented for a mortar reinforced by a textile fiber (TF). The development of molds in the culture environment of TF mortar was visible since the first week. In the containers with *A. niger* and *P. funiculosum*, it was observed that both molds developed, with a greater prominence of *A. niger*, to the detriment of *P. funiculosum*. In the containers with *, the presence of other molds was noticed [229].

Weekly, during the test period, the degree of mold development on the surface of the specimens was evaluated and this is summarized in Table 7.9. As presented for

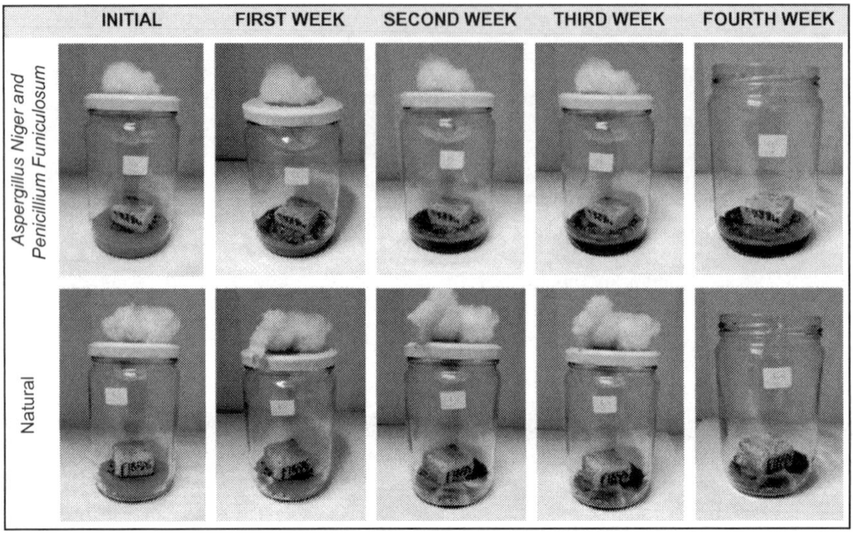

Fig. 7.40 Biocolonization development of a mortar with textile fibers waste [218] [229].

the previous wastes, in the second week of test, the REF mortar already presented some mold development in *natural inoculum*, until 20% of development (grades 1–2) [229]. The mortar with textile fibers, on the contrary, did not present any mold development for both sets of specimens analyzed (Fig. 7.41). Thus, it was concluded that the incorporation of these textile fibers, composed of acrylic threads, did not affect mold colonization.

Fig. 7.41 Biocolonization development in textile fiber–reinforced mortar surface [218] [229].

Throughout the biological colonization susceptibility test, the development of molds on the surface of the mortars was found to be relatively limited, although it had greater expression in the mortar with GFRP. It was also found that the *natural inoculum* of the samples led to a higher degree of contamination. Sterilization of the test specimens (performed according to the test standard ASTM D5590 [91]) may contribute to the discrepancy of results, as part of the fungi favoring nutrients can be eliminated. However, this sterilization and subsequent inoculation with a suspension of identified and registered fungi spores is critical so that it is possible to compare different materials within the same test by eliminating the variability resulting from the use of *natural inoculum*. This research identifies the need of validating in practice the importance of molds [218] [229], *A. niger* and *P. funiculosum* or others to be defined, recognizing the current technical developments in the area and the practice of colonization in mortars in general, as this is in continuous modification and depends on the location of the elements, as well as the climatic conditions at the moment [229].

7.4 Chapter conclusions

During their lifetime, renders are submitted to external actions that can affect their durability. In this chapter, some characteristics of the mortars over time or after ageing were discussed. It was noticed that wastes' incorporation can also contribute to modifications of mortars' characteristics and their performance when submitted to several actions. In a render, proper compatibility between the mortars and their substrate is essential. How this compatibility can affect mortars' durability was an issue discussed in this chapter.

In terms of deformability, which was analyzed through the ME determined by two different tests, it was noticed that mortars with or without wastes' incorporation modified these characteristics over time. In the first set of tests, mortars developed in laboratory and molded in prismatic specimens, kept in controlled temperature and humidity conditions, were tested over time. It was noticed that most of the mortars reduced their ME over time, improving their deformability. In the second set of tests, mortars were applied over a ceramic substrate and submitted to artificial ageing. In this set of tests, some mortars presented higher ME after ageing and others decreased their modulus, in comparison with the one before ageing. Glass was the waste that most changed mortars' characteristics with ageing. The ME is a first approach to the mortars' mechanical strength. Thus, a modification is usually related also with a change in mechanical behavior. In this sense, some tests about adherence strength and impact resistance test before and after artificial ageing were presented in this chapter. Some wastes improved the adherence strength of the mortars even after being submitted to artificial ageing, as well as the impact resistance, presenting mechanical strength greater than that of a REF mortar (without wastes). Concerning water permeability, it is also modified over time. Mortars due to internal microcrack, to external cracking or lack of adherence, generally increase their permeability over time. In the mortars with wastes' incorporation, a similar trend was noticed. Of the wastes exemplified in this chapter, GFRP waste was the one that presented the greatest increments of water permeability with ageing. This type of waste, although encapsulated in a cement matrix, due to its composition (polymeric resin), was deteriorated by ageing tests.

In this chapter, the influence of the wastes on mortars' esthetical appearance with ageing was also discussed. Generally, mortars loose some upper layer thickness over time, namely due to the water leaching. Cracking can also affect the general esthetical appearance of buildings. GFRP waste was the only one, from the set of examples, that cracked when submitted to artificial ageing. In addition to the loss of functional requirements, it also affects the mortars' esthetical appearance. At last, in this chapter, the increase of mortars' susceptibility to biocolonization with the wastes' incorporation was also discussed. However, of the wastes analyzed, it was found that the influence of this type of materials on this susceptibility is relatively limited.

Life cycle assessment

8

The concept of "sustainable construction" represents a paradigm for the different players in construction [232]. Although construction designers are concerned with choosing sustainable materials and solutions, construction companies are focused on maximizing the (economic) sustainability of their field of activity, by finding similar solutions or materials at a lower cost. Investors search for additional ways to promote their investment, decreasing at the same time their environmental impacts (EIs) [232]. However, it is necessary to define the concept of "sustainable construction" for all the actors. An option is by measuring the real sustainability of each project using a scientific-based tool.

Life cycle assessment (LCA) is a possible tool to measure the real environmental sustainability of a construction [5]. This tool allows evaluating and quantifying the environmental aspects and potential impacts associated to a product, process, or service [5] [233] [234] [235]. For that purpose, this methodology is divided in four phases [235] [236]: goal and scope definition, inventory analysis, impact assessment, and interpretation.

At first, the *goal and scope definition* of the assessment should be defined. This comprises the definition and description of the product, process, or activity, establishing the context, identifying the assessment boundaries and the environmental aspects considered [5] [235] [236].

A product passes through several activities during its lifetime, namely: production, usage, maintenance, and final disposal [235]. The combination of these activities is designated product life cycle. The LCA of a construction product can scope more or less activities and be more or less comprehensive, depending on the desired level of detail [5]. This level of detail is designated LCA boundary. There are three possible boundaries: "cradle to gate," "gate to grave," and "cradle to cradle" (Table 8.1). In the "cradle to gate" boundary, the LCA only considers the production stage. In the "gate to grave" boundary, the construction process, use, and end-of-life phases are considered. The "cradle to cradle" boundary comprises the previous two analyses and in addition the reuse, recover, or recycling options at the end-of-life of the materials, being the most complete boundary.

After the goal and scope definition, the following phase is the *inventory analysis*. Life cycle inventory is a process of LCA that quantifies the energy and raw material requirements, atmospheric and waterborne emissions, solid wastes, and other releases for the entire life cycle of a product, process, or activity [5] [235]. This phase involves the collection of data and calculation procedures to quantify the relevant inputs and outputs of a product, process, or activity [236]. The inventory can be based on environmental product declarations (EPDs), collected site-specific data, scientific studies, or international databases [5].

After the inventory, the following phase of LCA is the *impact assessment*. The life cycle impact assessment is the quantification of the potential human health and EIs of the environmental resources and releases identified in the life cycle inventory [5] [235].

Eco-efficient Rendering Mortars. DOI: 10.1016/C2018-0-04147-X

Table 8.1 Detailed life cycle stages of building materials based on European Standards. (Adapted from Silvestre [232]).

Life cycle assessment boundaries		**Life cycle stages**	**Life cycle stages designation and description**	
Cradle to cradle	Cradle to gate	Product stage (A1–A3)	A1	Raw material extraction and processing, processing of secondary material input
			A2	Transport to the factory
			A3	Manufacturing
	Gate to grave	Construction process stage (A4–A5)	A4	Transport to the building site
			A5	Installation into the building
		Use stage—information modules related to the building process (B1–B5)	B1	Use or application of the installed product
			B2	Maintenance
			B3	Repair
			B4	Replacement
			B5	Refurbishment
		Use stage—information modules related to the operation of the building (B6–B7)	B6	Operational energy use
			B7	Operational water use
		End-of-life stage (C1–C4)	C1	De-construction, demolition
			C2	Transport to waste processing
			C3	Waste processing for reuse, recovery and/or recycling
			C4	Disposal
		Benefits and loads beyond the system boundary (D)	D	Reuse, recovery and/or recycling potentials

Then, the results obtained should be interpreted and that is the last phase of the LCA. In this *interpretation* phase, the results obtained are assessed and a preferred product, process, or service can be selected, with a clear understanding of the uncertainty and the assumptions used to generate the results [5] [235] [236].

8.1 Application to mortars

The life cycle assessment (LCA) can be performed to products, processes, or services. In this book, the methodology of an LCA applied to render mortars is presented, more specifically to mortars with wastes incorporation in comparison with mortars without wastes. These wastes replace the initial constituents of the mortars: sand and cement. The incorporation of wastes in mortars is usually presented as an ecofriendly solution as they replace the initial constituents of the mortars and are simultaneously encapsulated, avoiding landfill deposition. In this book, a methodology to evaluate the mortars EIs in the product stage (cradle-to-gate boundary) is presented. From this methodology, it will be possible to state that mortars with wastes incorporation are in fact ecofriendly solutions not only as a means of waste management but also by avoiding the consumption of natural resources.

8.2 Case studies

In this section, the LCA of 11 mortars with wastes incorporation is presented. These wastes replaced sand or cement according to the information summarized in Table 8.2. In 9 of these modified mortars, 7 have wastes that replaced sand (natural aggregate). The replacement is higher in the aggregate incorporations as in the case

Table 8.2 Mortars with wastes incorporation composition (in volume), by several authors.

Type	Waste	Constituent replaced	Constituents (in volume)				References (based on)
			Sand	*Cement*	*Admixtures*	*Waste*	
Aggregate	Red ceramics	Sand	40%	19.8%	0.2%	40%	[221]
	Sanitary ware		-	19.8%	0.2%	80%	[5] [218]
	Construction demolition waste (CDW)—predominantly concrete		-	19.8%	0.2%	80%	[155]
	Construction demolition waste (CDW)		64%	19.8%	0.2%	16%	[155]
Filler	Red ceramics	Sand	72%	19.8%	0.2%	8%	[49]
	Sanitary ware		64%	19.8%	0.2%	16%	[5] [218]
	Construction demolition waste (CDW)—predominantly concrete		64%	19.8%	0.2%	16%	[154]
	Construction demolition waste (CDW)		68%	19.8%	0.2%	12%	[154]
	Glass fiber reinforced polymer (GFRP)		40%	19.8%	0.2%	40%	[5] [218]
Binder	Forest biomass bottom ashes	Cement	80%	16.8%	0.2%	3%	[5] [218]
Fibers	Textile fibers	All	76%	18.8%	0.2%	5%	[5] [218]

of sanitary ware and construction and demolition waste (CDW), mainly concrete, that replaced sand at 100%, meaning that the modified mortars contain no natural aggregate. In the filler incorporation, the volume of wastes is lower and varies from 10% (in the case of red ceramics) to 50% (in case of glass fiber reinforced polymer (GFRP)). The fillers of sanitary ware and of the two types of CDW replaced the natural aggregate at 15–20%. Also, in the case studies, there is also a modified mortar in which the cement was replaced by a waste: forest biomass bottom ashes. In this mortar, the initial binder was reduced by 15% by this waste. Finally, a mortar reinforced by textile fibers is also presented. These fibers were incorporated as an addition replacing part of the sand and cement.

All the mortars have a low binder: aggregate ratio. As noticed in the table, the volume of cement varies from 16.8% to 19.8%. This is a characteristic of render mortars. This type of mortars, as discussed throughout this book, is poor in cement. In this sense, the EIs considered below refer to render mortars. Mortars richer in cement, as masonry or reinforcement mortars, present more impacts than this type of mortars due to the volume of cement incorporated. That is because, as discussed later, cement is the constituent that contributes the most to the EIs of almost all types analyzed.

8.3 Methods adopted in mortars' case studies

8.3.1 Goal and scope definition

The aim of this assessment is to discuss the environmental benefits of mortars with wastes incorporation in several environmental characteristics. For that purpose, the LCA of the mortars presented in Section 8.3. was carried out based on ISO 14040 [236], ISO 14044 [237] and on EN 15804 [238].

The LCA includes the raw materials impacts, their transportation to the factory and the mortars' production, and is a cradle-to-gate analysis.

In the case studies, raw materials (sand, cement, wastes, and admixtures) are transported from their origin to the factory. Reception of these materials is the starting point. In general, raw materials are stored in silos (by gravity) upon arrival at the factory. Wastes that need any treatment process are only stored after passing through that process. Wastes as red ceramics, sanitary ware, or CDWs have to be crushed and sieved when arriving at the factory, as they do not have the adequate size to be incorporated in render mortars. Wastes are stored only after that treatment process. In the case studies, it was considered that the raw materials arrive in bulk by lorry and are discharged by gravity into silos, remaining there until mixing. In that phase, raw materials are mechanically weighted and mixed industrially until the mixture is homogeneous and then they are contained in Kraft paper bags (with a low-density polyethylene layer). After bagging, premixed mortars are stacked in pallets with a plastic protection (Polyethylene shrink film layers on the lateral and top of the pallet) and stored in a specific area inside the factory to avoid humidity. During the production process, some electric equipment is used as conveyor belt transport, automatic balance, filling machine, or automatic pallet wrapping. The impacts of these materials are also considered in the LCA.

8.3.2 Declared unit and system boundary

In the mortars' case studies, the functional unit considered by the authors was a volume unit of 1 m^3 of finished powder mortars [5] [218]. A "cradle to gate" boundary was considered in the case studies. In Fig. 8.1, this boundary is illustrated, and it includes the following:

- A1—Raw materials extraction and processing;
- A2—Raw materials transportation;
- A3—Manufacturing process that is divided into:
  - A3.1—packaging materials that leave the factory with the final product (raw materials and transport of the packaging materials);
  - A3.2—manufacturing process (energy used, internal transport, and waste production during the manufacturing process);
  - A3.3—production and disposal of wastes resulting from raw materials' packaging and from packaging materials' wastes.

8.3.3 EI categories analyzed

There are several EIs that can be considered in an LCA. In the case studies presented in this book, the LCA was assessed through eight EI categories that are required for EPDs by EN 15804 [238]. The EI studied by the authors are as follows:

- Abiotic depletion potential (ADP); two components: ADP–materials and ADP–fossil fuels;
- Global warming potential (GWP);
- Ozone depletion potential (ODP);
- Photochemical ozone creation potential (POCP);
- Acidification potential (AP) of land and water;
- Eutrophication potential (EP);
- Use of non-renewable primary energy resources (PE-NRe);
- Use of renewable primary energy resources (PE-Re).

The initial six impact categories listed were analyzed by the authors [5] [218] according to the Institute of Environmental Sciences (CML) from Leiden University, Netherlands, which according to the authors is one of the methods most applied in practice [5] [218] [233]. The EIs of the use of nonrenewable and renewable primary energy resources, the last two categories listed, were analyzed according to another method: cumulative energy demand (CED) Method.

8.3.4 LCA assumptions and limitations

The LCA of the case studies was based on some assumptions that led to limitations in the environmental assessment. The assumptions and corresponding limitations identified by the authors were [5] [218] as follows:

- The eight EI categories analyzed in the LCA are the ones that are most commonly used in civil engineering activities. Thus, there are other EIs that could be analyzed instead of these;
- The LCA boundary analyzed by the studies was cradle-to-gate (Fig. 8.1). Thus, the EIs of construction, use and end-of-life stages were not taken into account;

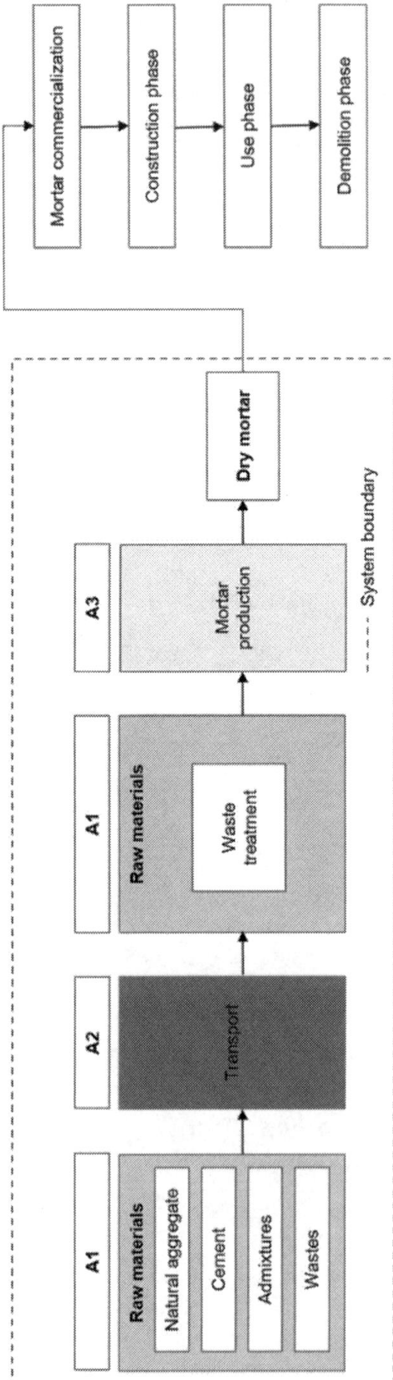

Fig. 8.1 LCA boundary.

- The data used to estimate the EIs of wastes processing (crushing and internal transport from crushing area to storing area) were collected from Portuguese companies that produce stone aggregates (granite and limestone). It was considered that the energy used to crush wastes was similar to the one used to crush granite and limestone, which can differ in practice;
- The EIs of the mortars production were estimated according to data collected from a specific Portuguese company. This is a limitation as the energy used can differ if a different company is chosen;
- This LCA was designed for a specific location of the factory, which affected the transportation distances and consequently the EIs related to the transport. Thus, a different location would be responsible for distinct transport distances and as a consequence would affect the overall results obtained in these LCAs;
- Mortars' bulk density was the only characteristic that was changed in the mortars' production EIs and other aspects could change these impacts;
- In some cases, the authors to estimate the EIs (of a product or process), were not able to use an EPD, a site-specific data collection or a scientific research. Whenever it occurred, the authors used data from databases from SimaPro software (Ecoinvent 3 and European Life Cycle Database (ELCD));
- Some wastes needed a simple treatment process before their application in mortars and it was not considered as the energy spent in that process was negligible in comparison to the other processes under analysis. Textile fibers [5] [218] were one of those wastes. They have to be cut before their addition in mortars, but the EIs of the cutting were not considered.

8.4 Life cycle inventory of mortars' case studies

To obtain the EIs inventories of raw materials, transportation, or mortars production, several sources and data locations had to be researched. In Table 8.3, this information is summarized.

The EIs of cement, air entrainer, superplasticiser, and water proofing admixtures were obtained from International EPDs. The ones of cement were obtained from a report of the European Cement Research Academy [4] and the admixtures from three different reports of the European Federation of Concrete Admixtures Associations [240] [241] [242], all from the year of 2015. The impacts of the use of sand in mortars were collected by a generic dataset available in SimaPro (using version 3.05 of CML 2 baseline 2000 released by CML in April 2013, and CED version 1.10).

Concerning wastes, and unless these need to pass through a transformation process before their incorporation in mortars, no impact from their manufacturing is considered because these production impacts are already associated to the product stage of the materials that originated them. Thus, from the wastes that are being analyzed in this book, GFRP powder, forest biomass ashes, or textile fibers have no associated impact before their incorporation in mortars. The remaining wastes considered have to be crushed and sieved beforehand and this impact was calculated and associated to the product stage "Raw materials" (A1). The LCA of these materials was obtained using site-specific data from a research study of Braga et al. [243] in a Portuguese company.

The impacts of raw materials transportation to the factory (A2) were estimated according to data from ELCD from the SimaPro software.

Table 8.3 Source and location of the databases used to estimate the environmental impacts of raw materials, transport and mortars' production.

Product stage	Raw materials	Technical details	Location	Sources
A1	Cement	CEM II	EU	International EPD system (ECRA [4])
	Sand	River sand	EU/PT	European Life Cycle Database (ELCD [239])
	Air entrainer	Powder admixture	EU	International EPD system (EFCA [240])
	Superplasticiser	Powder admixture	EU	International EPD system (EFCA [241])
	Water proofing	Powder admixture	EU	International EPD system (EFCA [242])
	Red ceramics	Aggregate waste (crushed and sieved)	EU/PT	Site-specific data (adapted from Braga et al. [243])
	Sanitary ware	Aggregate waste (crushed and sieved)	EU/PT	Site-specific data (adapted from Braga et al. [243])
	Construction demolition waste predominantly concrete (CDW-C)	Aggregate waste (crushed and sieved)	EU/PT	Site-specific data (adapted from Braga et al. [243])
	Construction demolition waste (CDW)	Aggregate waste (crushed and sieved)	EU/PT	Site-specific data (adapted from Braga et al. [243])
	Red ceramics	Powder waste (crushed and sieved)	EU/PT	Site-specific data (adapted from Braga et al. [243])
	Sanitary ware	Powder waste (crushed and sieved)	EU/PT	Site-specific data (adapted from Braga et al. [243])
	Construction demolition waste predominantly concrete (CDW-C)	Powder waste (crushed and sieved)	EU/PT	Site-specific data (adapted from Braga et al. [243])
	Construction demolition waste (CDW)	Powder waste (crushed and sieved)	EU/PT	Site-specific data (adapted from Braga et al. [243])
	Glass fiber reinforced polymer (GFRP)	Powder waste	EU/PT	None impact considered for raw materials
	Forest biomass ashes	Powder waste	EU/PT	None impact considered for raw materials
	Textile fibers	Fiber waste	EU/PT	None impact considered for raw materials
A2	Articulated transport		EU	European Life Cycle Database (ELCD [239])
A3	Powder mortars packaging (A3.1)		EU/PT	Site-specific data (adapted from Silvestre [232])
	Mortars manufacturing (A3.2)		EU/PT	Site-specific data (adapted from Silvestre [232])
	Production and disposal of packaging waste (A3.3)		EU/PT	Site-specific data (adapted from Silvestre [232])

Abbreviations: EU, European Union, PT, Portugal.

Finally, the impacts of mortars' production (A3) were adapted from a site-specific data from a research study by Silvestre [232] in Portuguese companies.

8.4.1 Raw materials (A1)

The render mortars considered in the case studied are composed of a natural aggregate, namely sand, cement, admixtures (air entrainer, superplasticiser, and waterproofing admixtures), and wastes, replacing the initial constituents as indicated in Table 8.2.

The impacts of these raw materials are considered in life stage A1. The EIs of cement, sand, and admixtures per kilogram are summarized in Table 8.4. Cement's EIs were collected from an EPD of the European Cement Research Academy [4]. According to the Farinha et al. [5], this EPD was chosen because it is the most recent average dataset for Europe for CEM II. The impacts of sand were estimated from the ELCD denominated "Sand 0/2 mm, wet and dry quarry, production mix, at plant, undried RER S." The EIs of admixtures were obtained from three EPDs of the European Federation of concrete admixtures associations [240] [241] [242].

As mentioned, wastes came without any environmental charge, meaning that wastes do not present any impact as raw materials. The impact of these wastes was considered in the industry that gave them origin. The only impacts that the wastes result from any treatment process they may be subjected to. Thus, in the case studies, some wastes do not present any impact in the raw materials stage, namely: GFRP or forest biomass ashes. The cutting process of textile fibers was not considered in this LCA and, in this way, there were also no EI associated to this process [5] [218]. The remaining wastes from the case studies have impacts associated to the treatment process, namely due to crushing and sieving. The impacts of these wastes are summarized in Table 8.5.

Red ceramics, sanitary ware, CDWs (mixed or mainly composed by on concrete) have to be crushed and sieved to be incorporated in mortars as aggregates or fillers, as they do not come from their industries in the suitable size to be directly incorporated. This treatment processes leads to EIs. The case studies adapted the data collected in a site-specific from a research study by Braga et al. [243] in Portuguese companies. The authors analyzed the EIs of the crushing and sieving limestone and granite aggregates process. In these case studies, it was considered that red ceramics and CDWs (composed by a mixture of wastes where red ceramics are included) had hardness similar to a weak limestone rock. On the other hand, CDWs mainly composed of concrete fragments have hardness closer to a compact limestone rock, and sanitary ware hardness more similar to granite aggregates. In this sense, the EIs were estimated according to the type of aggregates closer to the hardness of the wastes presented. These wastes were crushed and sieved in factory, as described in Section 8.3.1., and therefore it was also necessary to estimate the impacts of transportation inside the factory. These impacts were adapted from a previous research by Silvestre [232].

8.4.2 Transport (A2)

Transport impacts were estimated according to data from ELCD from SimaPro software, an "Articulated lorry transport" (with a maximum capacity of 27 tons) was considered in the case studies to transport the raw materials from their origin to the

Table 8.4 Environmental impact of cement, sand, and admixtures (for 1 kg).

Environmental impacts of the mortars' constituents		**Cement**	**Sand**	**Air entrainer**	**Superplasticiser**	**Water proofing**
Baseline CML method	ADP (kg Sb-eq.)	1.10×10^{-06}	7.31×10^{-11}	3.66×10^{-07}	1.10×10^{-06}	7.20×10^{-05}
	ADP (MJ)	2.84×10^{-03}	2.88×10^{-02}	$1.33 \times 10^{+01}$	$2.91 \times 10^{+01}$	$5.70 \times 10^{+01}$
	GWP (kg CO_2-eq.)	7.38×10^{-01}	2.46×10^{-03}	5.27×10^{-01}	$1.88 \times 10^{+00}$	$2.67 \times 10^{+00}$
	ODP (kg CFC11-eq.)	1.10×10^{-10}	3.83×10^{-10}	7.56×10^{-11}	2.30×10^{-10}	4.26×10^{-10}
	POCP (kg ethene-eq.)	1.17×10^{-04}	1.08×10^{-06}	3.45×10^{-04}	3.12×10^{-04}	9.62×10^{-04}
	AP (kg SO_2-eq.)	1.21×10^{-03}	1.90×10^{-05}	1.30×10^{-03}	2.92×10^{-03}	8.29×10^{-03}
	EP (kg PO_4^{-3}-eq.)	1.74×10^{-04}	2.03×10^{-06}	1.43×10^{-04}	1.03×10^{-03}	9.55×10^{-04}
Cumulative energy	PE–NRe (MJ)	3.04×10^{00}	4.23×10^{-02}	$1.42 \times 10^{+01}$	$3.14 \times 10^{+01}$	$6.08 \times 10^{+01}$
	PE–Re (MJ)	1.80×10^{-01}	1.70×10^{-03}	8.24×10^{-01}	$1.51 \times 10^{+00}$	$7.23 \times 10^{+00}$
References		ECRA [4]	ELCD [239]	EFCA [240]	EFCA [241]	EFCA [242]

Table 8.5 Environmental impact of several wastes that need to be crushed and sieved beforehand (for 1 kg).

Environmental impacts of the wastes		Aggregate wastes				Filler wastes			
		Red ceramics	*Sanitary ware*	*CDW*	*CDW-C*	*Red ceramics*	*Sanitary ware*	*CDW*	*CDW-C*
Baseline CML method	ADP (kg Sb-eq.)	2.48×10^{-10}	5.36×10^{-10}	2.48×10^{-10}	8.17×10^{-10}	2.48×10^{-10}	5.36×10^{-10}	2.48×10^{-10}	8.17×10^{-10}
	ADP (MJ)	4.10×10^{-03}	8.88×10^{-03}	4.10×10^{-03}	1.35×10^{-02}	4.10×10^{-03}	8.88×10^{-03}	4.10×10^{-03}	1.35×10^{-02}
	GWP (kg CO_2–eq.)	4.39×10^{-04}	9.50×10^{-04}	4.39×10^{-04}	1.45×10^{-03}	4.39×10^{-04}	9.50×10^{-04}	4.39×10^{-04}	1.45×10^{-03}
	ODP (kg CFC11–eq.)	1.52×10^{-11}	3.29×10^{-11}	1.52×10^{-11}	5.01×10^{-11}	1.52×10^{-11}	3.29×10^{-11}	1.52×10^{-11}	5.01×10^{-11}
	POCP (kg ethene–eq.)	1.30×10^{-07}	2.81×10^{-07}	1.30×10^{-07}	4.27×10^{-07}	1.30×10^{-07}	2.81×10^{-07}	1.30×10^{-07}	4.27×10^{-07}
	AP (kg SO_2–eq.)	3.38×10^{-06}	7.31×10^{-06}	3.38×10^{-06}	1.11×10^{-05}	3.38×10^{-06}	7.31×10^{-06}	3.38×10^{-06}	1.11×10^{-05}
	EP (kg PO_4^{-3}–eq.)	6.02×10^{-07}	1.30×10^{-06}	6.02×10^{-07}	1.98×10^{-06}	6.02×10^{-07}	1.30×10^{-06}	6.02×10^{-07}	1.98×10^{-06}
Cumulative energy	PE-NRe (MJ)	5.02×10^{-03}	1.09×10^{-02}	5.02×10^{-03}	1.66×10^{-02}	5.02×10^{-03}	1.09×10^{-02}	5.02×10^{-03}	1.66×10^{-02}
	PE-Re (MJ)	3.27×10^{-03}	7.08×10^{-03}	3.27×10^{-03}	1.08×10^{-02}	3.27×10^{-03}	7.08×10^{-03}	3.27×10^{-03}	1.08×10^{-02}
References		Based on Braga et al. [243]							

Table 8.6 Environmental impact of the transport (1 kg.km) in Europe.

Environmental impacts		Transport	Reference
Baseline CML method	ADP (kg Sb-eq.)	1.98×10^{-12}	ELCD [239]
	ADP (MJ)	6.99×10^{-04}	
	GWP (kg CO_2-eq.)	4.99×10^{-05}	
	ODP (kg CFC11-eq.)	1.01×10^{-13}	
	POCP (kg ethene-eq.)	1.59×10^{-08}	
	AP (kg SO_2-eq.)	2.24×10^{-07}	
	EP (kg PO_4^{-3}-eq.)	5.11×10^{-08}	
Cumulative Energy	PE-NRe (MJ)	7.03×10^{-04}	
	PE-NR (MJ)	7.95×10^{-07}	

factory [5] [218]. It was considered that an empty lorry produces less EIs than when it is full [232] and, in this sense, in the lorry return trip the EIs were reduced by 30% [5] [218] [244]. In Table 8.6, the impacts of transportation in the environmental categories analyzed in the case studies are summarized.

Transport impacts (product stage—A2) depend directly on the distance between the raw materials and the factory. In Table 8.7, the distance between the waste's origin and the manufacturer (effective distance) and the design distance are summarized. The design distance is the equivalent one considering that the lorry has to return to the initial location after unloading the wastes in the factory. That return distance, as already mentioned, is reduced by 30%, meaning that designed distance is given by the multiplication of the effective distance by 1.7. In the case studies, the sands' transport distance was 107 km away to the factory, the cement's distance was about 28 km, and the distance from the admixtures store to the factory was 249 km. The wastes incorporated in the mortars come from different locations. Forest biomass ashes are the waste, from the case studies, more distant from the factory. The waste travels 500 km (design distance) from their origin to the factory.

8.4.3 Production process (A3)

The mortars' production impacts were also estimated in the case studies. This process was adapted from site-specific data from a previous research [232] by a Portuguese company. The impacts of the production process are given by a cubic meter of powder mortar and they are summarized in Table 8.8. Thus, the bulk density of the mortars is the main characteristic that affects the EIs of mortars' production.

8.5 LCA of mortars' case studies

8.5.1 Abiotic depletion potential

ADP is the first EI category presented. This category corresponds to mineral and resources used [232] and is, in this sense, mainly influenced by the rate of resources extracted [245]. The effect of this consumption on their depletion is estimated

Table 8.7 Transport and transportation distance from the waste origin to the factory.

Type	Raw materials	Constituent replaced	Transport characteristics			References (Based on)
			Type of transport	*Effective distance (km)*	*Design distance (km)*	
General constituents	Cement	-	Articulated lorry transport	28	48	[5] [218]
General constituents	Sand	-	Articulated lorry transport	107	182	[5] [218]
General constituents	Admixtures	-	Articulated lorry transport	249	423	[5] [218]
Aggregate wastes	Red ceramics	Sand	Articulated lorry transport	71	121	[221]
Aggregate wastes	Sanitary ware	Sand	Articulated lorry transport	166	282	[5] [218]
Aggregate wastes	Construction demolition waste (CDW)—predominantly concrete	Sand	Articulated lorry transport	222	377	[155]
Aggregate wastes	Construction demolition waste (CDW)	Sand	Articulated lorry transport	48	82	[155]
Filler wastes	Red ceramics	Sand	Articulated lorry transport	71	121	[49]
Filler wastes	Sanitary ware	Sand	Articulated lorry transport	166	282	[5] [218]
Filler wastes	Construction demolition waste (CDW)—predominantly concrete	Sand	Articulated lorry transport	222	377	[154]
Filler wastes	Construction demolition waste (CDW)	Sand	Articulated lorry transport	48	82	[154]
Filler wastes	Glass fiber reinforced polymer (GFRP)	Sand	Articulated lorry transport	58	99	[5] [218]
Binder wastes	Forest biomass ashes	Cement	Articulated lorry transport	294	500	[5] [218]
Fiber wastes	Textile fibers	All	Articulated lorry transport	24	41	[5] [218]

Table 8.8 Environmental impact of the production process of 1 m^3 of powder mortar.

Environmental impacts		Mortars' production (A3)			Reference
		A3.1	*A3.2*	*A3.3*	
Baseline CML method	ADP (kg Sb-eq.)	8.59×10^{-05}	1.05×10^{-09}	5.50×10^{-10}	[5] [218] [232]
	ADP (MJ)	$4.37 \times 10^{+02}$	2.51×10^{-02}	8.31×10^{-04}	
	GWP (kg CO_2-eq.)	$2.78 \times 10^{+01}$	2.16×10^{-03}	7.57×10^{-05}	
	ODP (kg CFC11-eq.)	1.75×10^{-06}	2.29×10^{-10}	8.95×10^{-12}	
	POCP (kg ethene-eq.)	1.11×10^{-02}	4.65×10^{-07}	2.07×10^{-08}	
	AP (kg SO_2-eq.)	1.40×10^{-01}	1.25×10^{-05}	2.68×10^{-07}	
	EP (kg PO_4^{-3}-eq.)	4.70×10^{-02}	2.54×10^{-06}	2.05×10^{-07}	
Cumulative energy	PE-NRe (MJ)	$5.10 \times 10^{+02}$	6.09×10^{-02}	1.23×10^{-03}	
	PE-NR (MJ)	$1.12 \times 10^{+03}$	6.53×10^{-03}	-6.37×10^{-05}	

according to their availability stock at a global scale [246]. This impact category is divided into two components: a material component and a fossil fuels component.

The effects of raw materials, transportation, and production of a standard render mortar (reference mortar) in abiotic depletion EI are represented in Fig. 8.2. In the first component (materials), it can be noticed that cement is the mainly responsible for the abiotic depletion impact, representing 76% of the total impact of the render. Admixtures are the raw materials that, after cement, more increase this impact, being responsible for 23% of it. Sand's impact, when compared with the remaining ones, does not contribute actively to ADP. In the energy component (second component of ADP), transportation was responsible for half of the total impact, admixtures for 27%, and sand and mortars production for 11% and 12%, respectively.

Wastes can reduce the EIs in abiotic depletion as they replace the initial constituents as natural aggregate or cement. In Figs. 8.3 and 8.4, some examples of render mortars with wastes incorporation and their impact in both components of ADP are presented. As far as materials' components are concerned, and as the cement is the mainly responsible for abiotic impact, the wastes that replaced part of this constituent are the ones that presented better behavior. In the examples shown in Fig. 8.3, the mortar with forest biomass ashes [218], which replaced cement by 15%, decreased the abiotic depletion by 11%. Other binder wastes with the same type of admixtures would also reduce this environmental category. Textile fibers, as an addition, also replaced a small amount of cement, which resulted in a reduction of ADP by 4%.

In the fossil fuel component of ADP, the transportation impact is the mainly responsible for abiotic depletion. This means that the distance between the raw materials extraction and the manufacturer is the main influencer of ADP fossil fuel

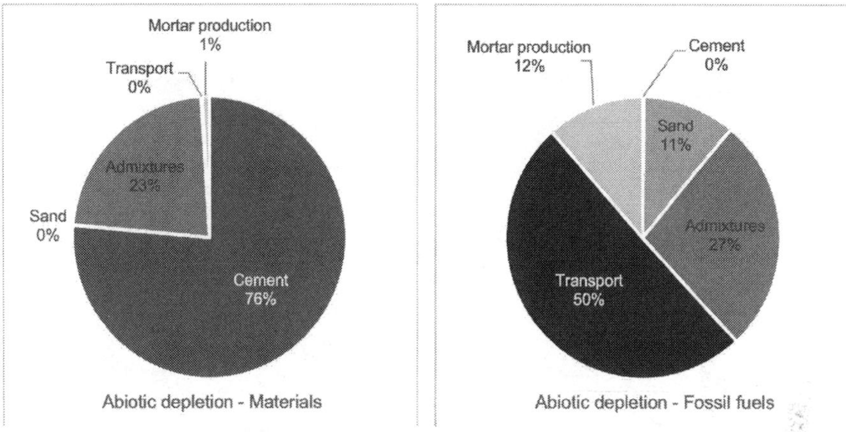

Fig. 8.2 Abiotic depletion potential—contribution of raw materials, transportation, and mortars production [5] [218].

components, because transport uses fossil fuels as energy. The wastes that came from a longer distance were CDW (mainly concrete) and forest biomass ashes, respectively, from a distance of 220 km and 300 km. CDW replaced the natural aggregate (at 100%) and came from a distance of 107 km, half the distance of CDW waste. Forest biomass ashes replaced cement (at 15%) and came from a shorter distance (about 30 km), meaning that the distance was 10 times lower. These distances were responsible for increments of ADP fossil fuels components. The mortar that incorporated CDW (mainly concrete) presented an impact 27% higher than the reference mortar. Forest

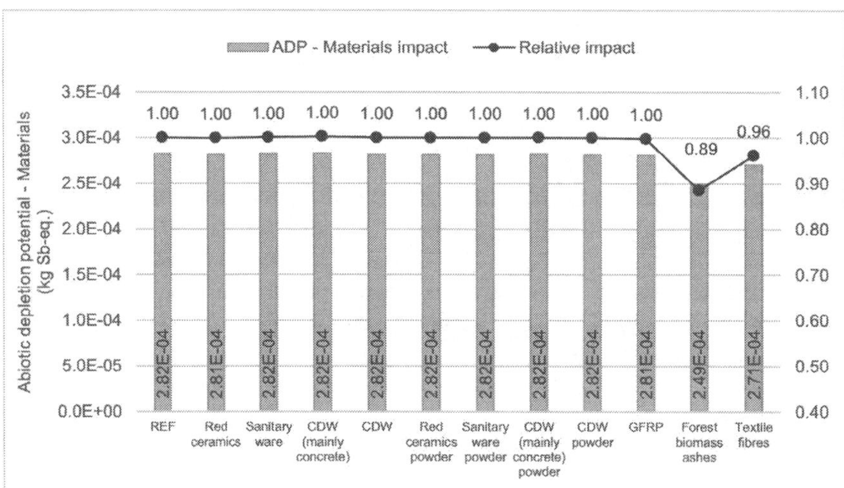

Fig. 8.3 Abiotic depletion potential—materials (kg Sb-eq.) [5] [49] [154] [155] [218] [221].

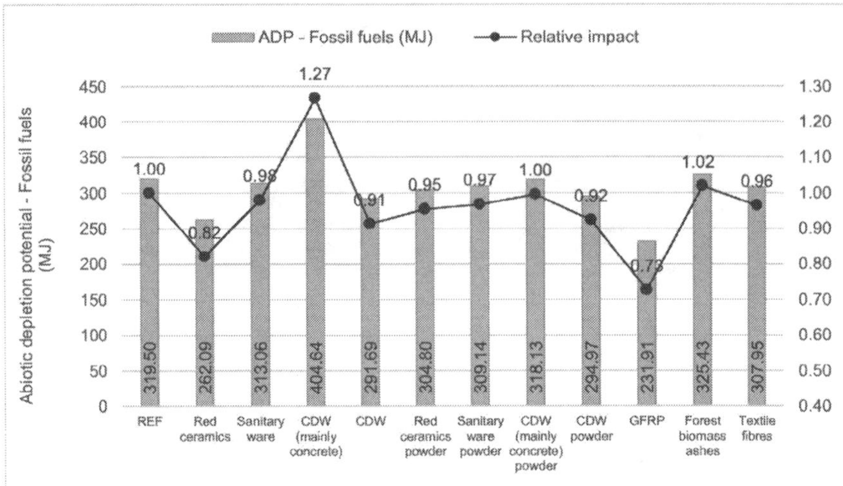

Fig. 8.4 Abiotic depletion potential—fossil fuels (kg MJ) [5] [49] [154] [155] [218] [221].

biomass ashes presented an impact of 2% in this ADP component. Although the distance of forest biomass ashes was higher than that of CDW (mainly concrete), the volume of incorporation was considerably lower, which resulted in a lower impact. When CDW (concrete) was incorporated as a powder waste, the impact was equal to the one of the reference mortar, that is about 318 MJ. That occurred also because the ratio of incorporation was lower, about 20%, while for aggregate incorporation the ratio was 100%, both replacing sand. The remaining mortars with wastes incorporation, represented in Fig. 8.4, presented a lower ADP impact than the reference mortar's. The mortar that presented lower impact was the one with GFRP waste, with an impact 27% lower than that of the reference mortar.

8.5.2 Global warming potential

GWP, known as "Greenhouse effect," is the EI responsible for the greenhouse gases' emission into the atmosphere [232] [245]. These gases increase the global medium temperature of the Earth [5] [232] as they reflect or trap part of the radiation emitted by the Earth's surface, such as the infrared reflection of solar radiation. The increment of temperature can result in several climate changes that can affect the ecosystems and human health [232]. According to the U.S. Environmental Protection Agency [235], several EIs are associated to GWP such as polar ice meltdown, soil moisture loss, longer seasons, forest loss/change, and change in wind and ocean patterns [5]. GWP impact is quantified by a kilogram of CO_2 equivalent.

Global warming impact in render mortars product (cradle to gate boundary) is mainly influenced by cement content. Fig. 8.5 shows an example of a reference

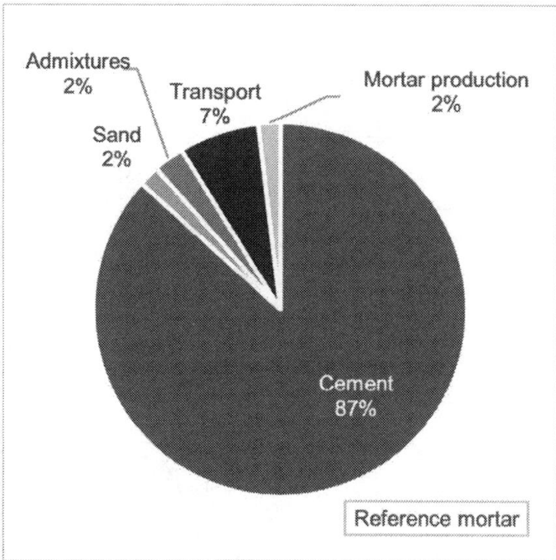

Fig. 8.5 Global warming potential—contribution of raw materials, transportation, and mortars production [5] [218].

mortar and the influence of raw materials, transportation of these materials until manufacturer, and mortars' production. In this example, cement is responsible for 87% of the total GWP impact.

The incorporation of binder wastes, that is wastes that replace cement, has a direct influence on the reduction of CO_2 emissions and consequently on the reduction of the GWP. In this sense, the incorporation of forest biomass ashes in render mortars, replacing cement, is responsible for a reduction of GWP. The replacement of cement at 15% of this waste reduced the GWP by 13% (Fig. 8.6). Some wastes as red ceramics (aggregate or powder particles), CDW (aggregate or powder particles), or GFRP powder produced decreases in GWP between 1% and 4%. These are associated to the reduction of sand content and to the transportation of the raw materials to the factory. CDW (mainly concrete) waste, due to the transport distance to the mortars' manufacturer, incremented the GWP impact by 4%. The transport considered in the life cycle inventory uses fossil fuels, being responsible for CO_2 emissions, which incremented the GWP impact.

8.5.3 Ozone depletion potential

ODP is the EI that represents the reduction of the ozone layer in the atmosphere. This layer is reduced with the concentration of halocarbons in the atmosphere [5] as these gases dissociate the ozone molecules presented, reducing this layer [245]. The

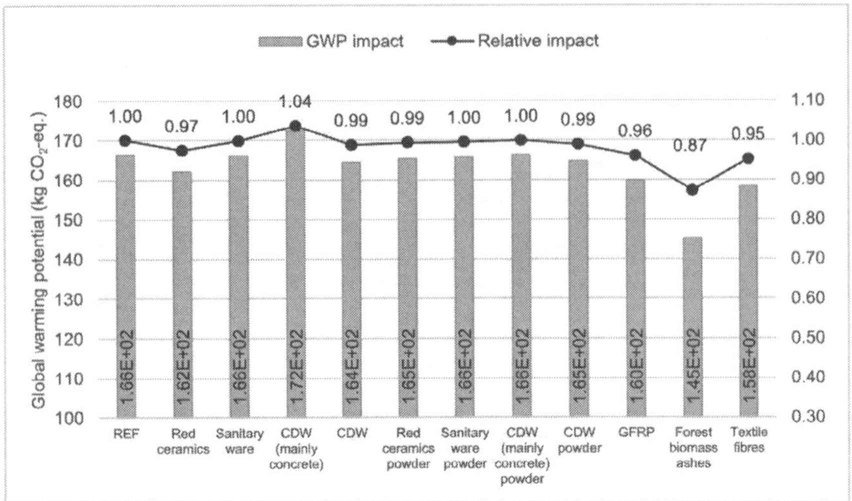

Fig. 8.6 Global warming potential (in kg CO_2 - eq.) [5] [49] [154] [155] [218] [221].

depletion of the stratospheric ozone layer increases the radiation that can reach the Earth's surface, such as the ultraviolet radiation [232] [245], which results in several climate changes that can affect the ecosystems and human health [5] [232]. ODP is evaluated by a kilogram of CFC-11 (Trichlorofluoromethane) equivalent.

In ODP impact, sand is the raw material that contributes the most to this impact. It is responsible for 55% of the total impact of the reference mortar, considered as a reference in Fig. 8.7. The mortars' production process is the component that comes after sands' impact and it is responsible for 40% of the ODP impact. The remaining 5% are distributed by transportation and cement.

In Fig. 8.8, the ODP impacts of 9 mortars that directly replaced sand by wastes are illustrated, namely: red ceramics, sanitary ware, CDW (mainly concrete) or CDW, all as aggregates or powders, and GFRP as powder waste. In all, the ODP impact was reduced. The main reductions can be noticed in the mortars with sanitary ware and CDW (mainly concrete) that replaced sand at 100%. These mortars decreased the ODP impact by 59% and 53%, respectively. A binder waste (forest biomass ashes) and the corresponding impact on ODP are also illustrated in Fig. 8.8. It also reduced the impact by 1%, which is not considerable as it did not significantly change the mortars' production or replaced sand. The incorporation of textile fibers as an addition of 5% also resulted in a decrease of the ODP impact by 5%.

8.5.4 Photochemical ozone creation potential

POCP is another EI that can be produced artificially, namely due to the photochemical oxidation of gases emitted by some vehicles [232]. This type of transport pro-

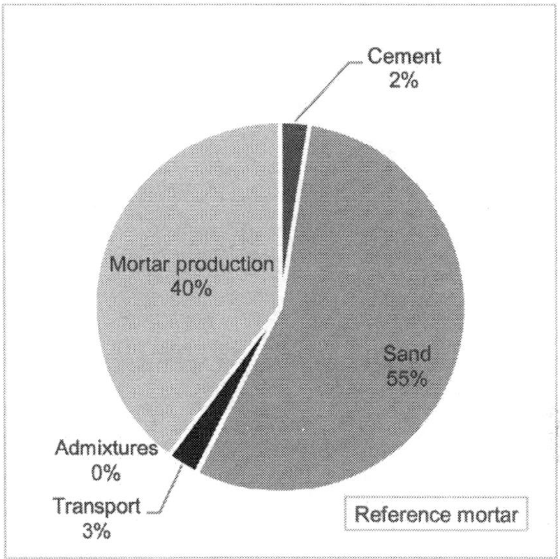

Fig. 8.7 Ozone depletion potential—contribution of raw materials, transportation, and mortars production [5] [218].

duces carbon monoxide and volatile organic compounds (VOC). When sunlight hits VOC, they form airborne particles and ground-level ozone commonly called smog [5] [232]. The concentrations of ozone are higher in the summer. Its toxic nature can directly affect human health [5], causing eye irritation, respiratory tract and lung

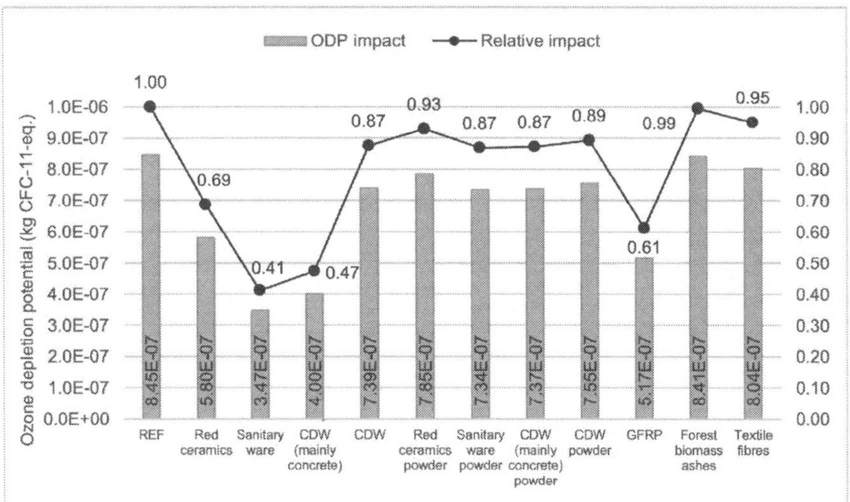

Fig. 8.8 Ozone depletion potential (in kg CFC-11 - eq.) [5] [49] [154] [155] [218] [221].

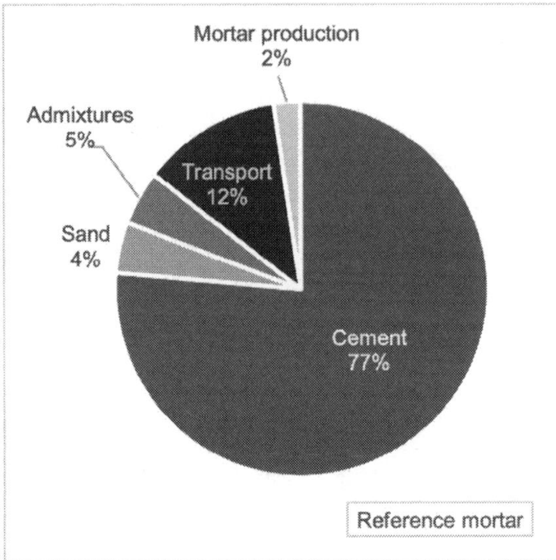

Fig. 8.9 Photochemical ozone creation potential—contribution of raw materials, transportation, and mortars production [5] [218].

irritation [235], and ecosystems disruption [232]. POCP is evaluated by a kilogram of ethane equivalent.

Fig. 8.9 presents the influence of raw materials (sand, cement, and admixtures), their transportation to the factory, and the mortars' production on the POCP impact of the final product for a render mortar. From the figure, it can be noticed that cement is the raw material that most contributes to the POCP impact, representing 77% of the total. Next is transport (A2) that most contributes to this impact. In this sense, the wastes that most reduce this impact are the ones that replaced cement or the ones that reduce the distance to the factory. Forest biomass ashes, replacing cement at 15%, are responsible for reducing the POCP impact by 11% (Fig. 8.10). The decrease of POCP could be higher if the distance between the waste and the factory were lower. Concerning CDW (mainly concrete), it was the waste that most incremented the POCP impact, by 5%. This increase is due to the distance between the waste and the factory. If the distance were similar to the one of sand, the POCP impact would be 5% lower than the one obtained by the reference mortar. Once more, the transportation distance proved to be an important factor in wastes' incorporation decision. The remaining wastes reduced the POCP impact by between 0% and 7%, depending on the wastes, as illustrated in Fig. 8.10. In addition to the forest biomass ashes, the waste that most reduced the POCP impact was GFRP. The reduction of the impact was due to the reduction of sand content (by 50%) and the lower distance to the factory than the natural aggregate.

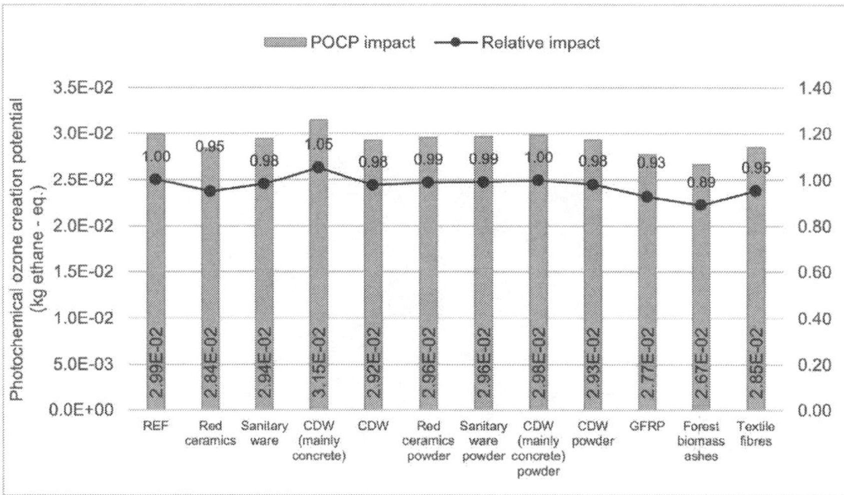

Fig. 8.10 Photochemical ozone creation potential (in kg ethane - eq.) [5] [49] [154] [155] [218] [221].

8.5.5 Acidification potential of land and water

AP is the EI that quantifies the amount of gases responsible for the acidification of soils, ground and surface water, effects on animals, ecosystems, and also the building environment [235]. The main gases responsible for acidification are sulfur dioxide (SO_2), nitrous oxides (NOx), and reduced nitrogen (NHx) [232] [245]. According to Farinha et al. [5], the sulfur compounds deposit within 2–4 days after emission and, therefore, affect areas close to the emission source [232]. AP impact is quantified by the amount of SO_2 equivalent per kilogram.

In Fig. 8.11, the influence of raw materials, transportation, and mortars' production impacts in AP of a reference render mortar is illustrated. Raw materials contribute 80% of the total AP impact, where cement contributes 70%, sand 7%, and admixtures 3%. The transportation of raw materials until the factory is responsible for 15% of the total AP impact and the mortars' production for the remaining 5%.

Wastes incorporated in mortars replacing raw materials, namely cement, can contribute to the reduction of land and water acidification, unless the distance to the factory makes the benefit of raw materials replacement unviable. In Fig. 8.12, 11 mortars with wastes incorporation are presented as examples. Ten of the 11 mortars illustrated have a lower AP impact than the reference mortar. The major decreases were noticed when forest biomass ashes replaced 15% of cement (by volume) and when GFRP replaced sand at 50% (by volume), both presenting a reduction of 10%. The reductions were due to the different types of replacement. On the contrary, CDW (mainly concrete) incremented the AP impact by 6%. This increment was related to

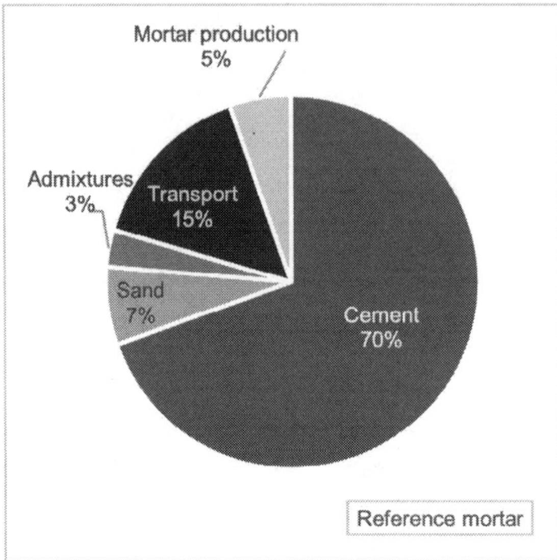

Fig. 8.11 Acidification potential—contribution of raw materials, transportation, and mortars production [5] [218].

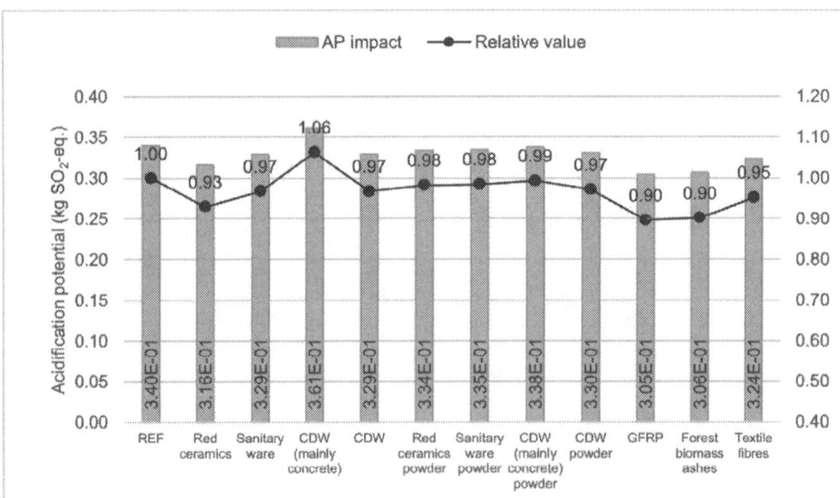

Fig. 8.12 Acidification potential (in kg SO_2 - eq.) [5] [49] [154] [155] [218] [221].

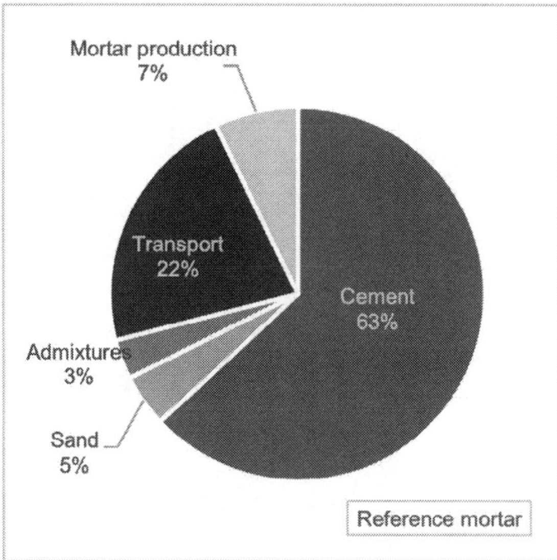

Fig. 8.13 Eutrophication potential—contribution of raw materials, transportation, and mortars production [5] [218].

the transportation distance betwee n the waste and the factory. If the distance travelled by CDW waste were similar to the sands', the waste not only would have reduced their AP impact but would also have an impact 7% lower than the one of the reference mortar, matching the impact of red ceramics (about 0.34 kg SO_2 eq.).

8.5.6 Eutrophication potential

EP is the EI responsible for the enrichment of nutrients in soil or water. This enrichment can be due to nitrogen and phosphorus from polluting emissions, wastewater, and fertilizers, originating excessive development of algae and plants [5] [245]. In water, this excessive development of microorganisms decreases the rates of oxygen and solar energy, leading to the contamination of plants and groundwater in terrestrial eutrophication [5] [235] [245].

Cement is the raw material that influences the EP impact in render mortars the most, in an analysis of cradle to gate boundary, as illustrated in Fig. 8.13. Cement is responsible for 63% of the total EP impact, sand by 5%, and admixtures by 3%. The transportation of raw materials to the factory, from the example presented is responsible for 22% of the EP total impact. The mortar production in factory represents the remaining 7% of EP total impact.

In Fig. 8.14, some examples of render mortars with wastes incorporation and their EP impact are represented. The impact of these modified mortars and relative to the

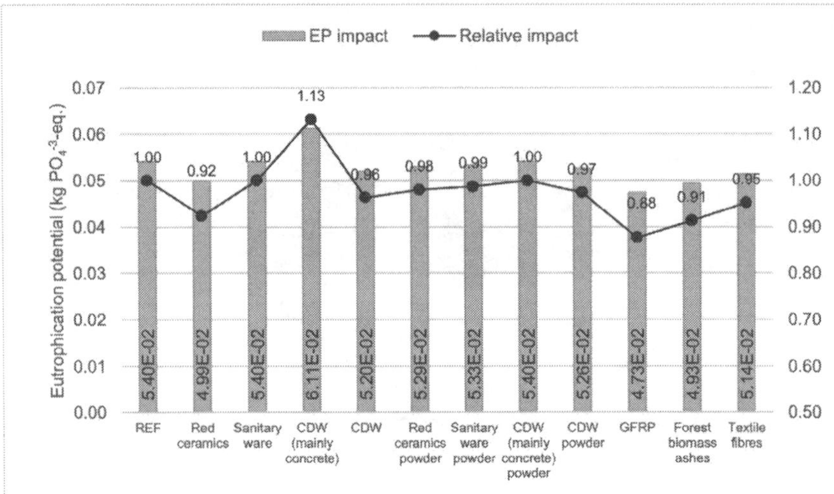

Fig. 8.14 Eutrophication potential (in kg PO_4^{-3} - eq.) [5] [49] [154] [155] [218] [221].

reference mortar is also illustrated. Eight of the 11 mortars with wastes represented had a lower EP impact than the conventional mortar, two had a similar impact, and one a higher impact. The mortars that presented lower impact were the ones that replaced sand at 50% with red ceramic wastes, the one that replaced sand at 50% of GFRP powder waste, and the one that replaced cement at 15% (forest biomass ashes), all by volume. The reductions of EP impact were 8%, 12%, and 9%, respectively. In the first two cases (red ceramics and GFRP powder), the reduction was due to the sand's replacement and to a lower distance between the wastes and the factory when compared with the distance from the sand source to the factory. In the case of forest biomass ashes, the reduction of EP impact was due to cement replacement. CDW (mainly concrete) was responsible for an increment in the EP impact. The mortar increased 13% more this impact than the reference mortar. This increment was due to the long distance that the waste travelled to the factory, as was already seen in other EI categories. The transportation distance and its associated impact, for some categories, are higher than the benefits that mortars would have had due to sand's replacement.

8.5.7 Use of nonrenewable primary energy resources

Nonrenewable primary energy (PE-NRe) is produced by resources that exist in a finite amount and cannot be replenished in a human time scale [5] [238].

The influence of raw materials, transportation, and production of render mortars on the use of nonrenewable primary energy resources is illustrated in Fig. 8.15, as an example. In this case, raw materials are responsible for 64% of the total PE-NRe impact, transportation for 17%, and production for 9%. Cement is the raw material that contributes the most to PE-NRe impact, being responsible for 60% of the total mortars' impact.

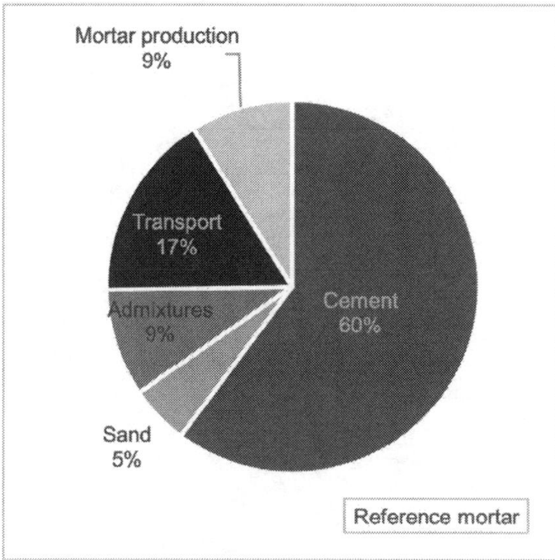

Fig. 8.15 Use of nonrenewable primary energy resources—contribution of raw materials, transportation, and mortars production [5] [218].

In Fig. 8.16, several examples of mortars with wastes incorporation and their PE-NRe impact are represented. As noticed, all mortars, with the exception of the one that contains CDW (mainly composed of concrete), presented a lower impact. The reduction of impact is mainly due to the replacement of the initial constituents (cement and sand). The major reduction of PE-NRe impact was noticed for the mortar with forest biomass ashes waste. That was because it was the only waste presented in the figure that replaced cement, the constituent that contributes the most to this EI. Forest biomass ashes reduced PE-NRe impact by 11%. On the contrary, CDW (mainly concrete) incremented this impact by 7%. Once more, the long distance of waste to factory was the main reason for this increment. The transportation distance is, in fact, a preponderant factor in LCA calculation.

8.5.8 Use of renewable primary energy resources

The renewable primary energy comes from renewable nonfossil sources, corresponding to resources that grow naturally, is replenished or naturally cleansed in a human time scale [5] [238].

In Fig. 8.17, an example of the impact of raw materials, transportation, and mortars' production in the use of renewable primary resources (PE-Re) impact is illustrated. The influence of cement in PE-Re impact or in PE-NRe is similar (around 60%); however, the mortars' production was incremented to 18% and admixtures to 15%, while sand was reduced to 4% and transportation represented 0% of this impact, due to the means of transport considered in this LCA.

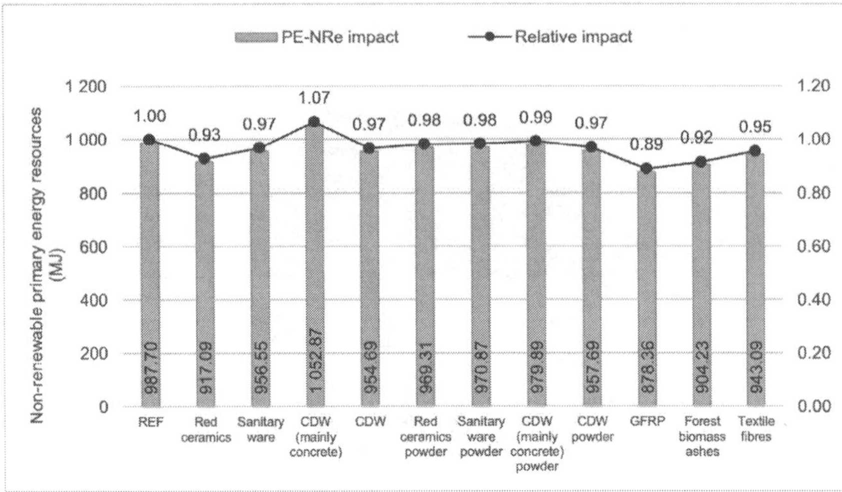

Fig. 8.16 Nonrenewable primary energy resources impact (in MJ) [5] [49] [154] [155] [218] [221].

The PE-Re impact of several mortars with wastes incorporation is illustrated in Fig. 8.18. The ones that replaced sand as aggregates were red ceramics, sanitary ware, CDW (mainly concrete), and CDW. All of these wastes passed through a treatment process before their incorporation in mortars. They had to be crushed and sieved beforehand. These treatments were responsible for incrementing the PE-Re impact.

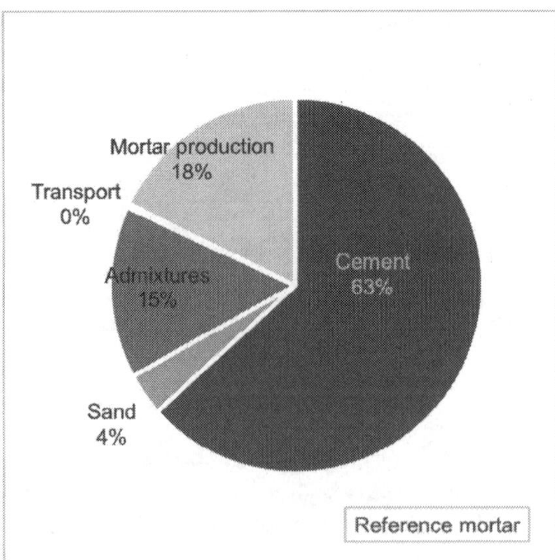

Fig. 8.17 Use of renewable primary energy resources—contribution of raw materials, transportation, and mortars production [5] [218].

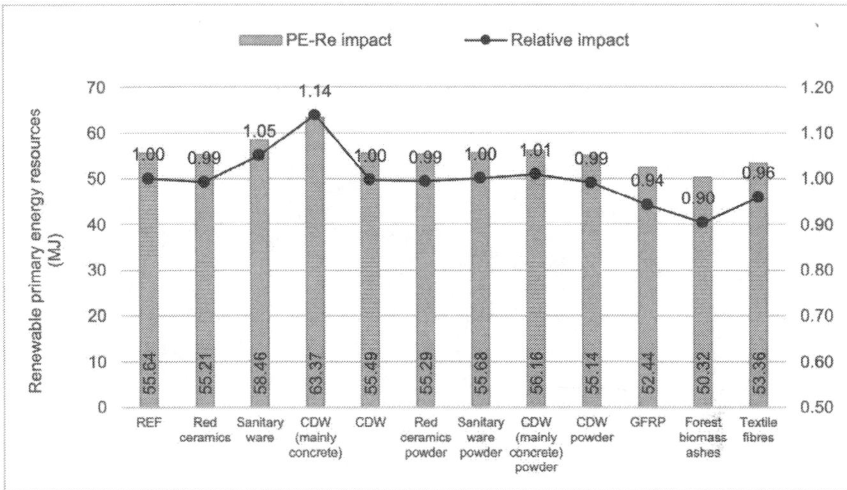

Fig. 8.18 Renewable primary energy resources impact (in MJ) [5] [49] [154] [155] [218] [221].

These increments were higher in the sanitary ware waste and in CDW (mainly concrete). These two wastes required more energy to be treated due to their hardness than the remaining ones. Red ceramics, even with the treatment process, presented an impact 1% lower than the reference mortar and CDW balanced the impact of the reference mortar. The powder wastes also passed through a treatment process, with the exception of the GFRP waste. The volume of powder wastes incorporation was lower than that of the aggregate wastes, which resulted in lower PE-Re impacts. Only CDW (mainly concrete) presented an impact of 1%, in comparison with the reference mortar. The remaining powder wastes presented an impact equal or lower than the reference. Forest biomass ashes reduced this impact by 10%, when replacing cement at 15%, being the waste that reduced this impact the most.

8.6 Chapter conclusions

Wastes incorporated in mortars are usually considered as eco-solutions, as they both replace conventional constituents (as natural aggregate and/or binder) and reduce waste landfill deposition. Sustainability and sustainable materials are, however, hard to define and to quantify. In this book, a methodology for a LCA was presented as a tool to quantify the sustainability of mortars with wastes incorporation. 11 case studies of mortars with distinct waste types (aggregates, fillers, binders and fibers) were presented as examples. In this methodology, the EIs of the mortars generated during the product stage were analyzed, corresponding to a cradle to gate boundary. The impacts analyzed were the ones that are usually considered in civil engineering activities: ADP, GWP, ODP, POCP, AP, EP, Use of non-renewable primary energy resources (PE-NRe), and use of renewable primary energy resources (PE-Re).

Table 8.9 Environmental impacts of mortars case studies for all categories.

Mortars		ADP	ADP	GWP	ODP	POCP	AP	EP	PE-NRe	PE-Re
		kg Sb eq.	*MJ*	*kg CO_2 eq.*	*kg CFC11 eq.*	*kg ethene eq.*	*kg SO_2 eq.*	*kg PO_4^{-3} eq.*	*MJ*	*MJ*
	Reference mortar	$2{,}82 \times 10^{-04}$	319.5	166.4	8.45×10^{-07}	2.99×10^{-02}	0.34	5.40×10^{-02}	987.7	55.6
Aggregate wastes	Red ceramics	$2{,}81 \times 10^{-04}$	262.1	162.1	5.80×10^{-07}	2.84×10^{-02}	0.32	4.99×10^{-02}	917.1	55.2
	Sanitary ware	$2{,}82 \times 10^{-04}$	313.1	166.1	3.47×10^{-07}	2.94×10^{-02}	0.33	5.40×10^{-02}	956.5	58.5
	CDW (mainly concrete)	$2{,}82 \times 10^{-04}$	404.6	172.4	4.00×10^{-07}	3.15×10^{-02}	0.36	6.11×10^{-02}	1052.9	63.4
	CDW	$2{,}82 \times 10^{-04}$	291.7	164.3	7.39×10^{-07}	2.92×10^{-02}	0.33	5.20×10^{-02}	954.7	55.5
Filler wastes	Red ceramics powder	$2{,}82 \times 10^{-04}$	304.8	165.3	7.85×10^{-07}	2.96×10^{-02}	0.33	5.29×10^{-02}	969.3	55.3
	Sanitary ware powder	$2{,}82 \times 10^{-04}$	309.1	165.6	7.34×10^{-07}	2.96×10^{-02}	0.33	5.33×10^{-02}	970.9	55.7
	CDW (mainly concrete) powder	$2{,}82 \times 10^{-04}$	318.1	166.2	7.37×10^{-07}	2.98×10^{-02}	0.34	5.40×10^{-02}	979.9	56.2
	CDW powder	$2{,}82 \times 10^{-04}$	295.0	164.6	7.55×10^{-07}	2.93×10^{-02}	0.33	5.26×10^{-02}	957.7	55.1
	GFRP	$2{,}81 \times 10^{-04}$	231.9	159.8	5.17×10^{-07}	2.77×10^{-02}	0.30	4.73×10^{-02}	878.4	52.4
Binder waste	Forest biomass ashes	$2{,}49 \times 10^{-04}$	325.4	145.2	8.41×10^{-07}	2.67×10^{-02}	0.31	4.93×10^{-02}	904.2	50.3
Fiber waste	Textile fibers	$2{,}71 \times 10^{-04}$	308.0	158.3	8.04×10^{-07}	2.85×10^{-02}	0.32	5.14×10^{-02}	943.1	53.4

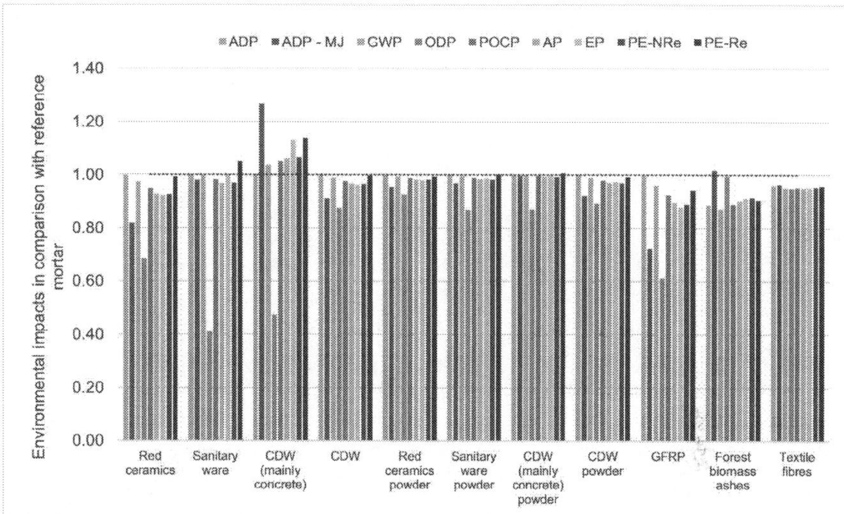

Fig. 8.19 Relative environmental impact of blended mortars relative to a reference mortar [5] [49] [154] [155] [218] [221].

In Table 8.9, the EIs of the case studies mortars obtained are summarized and, in Fig. 8.19, the relative impact of the modified mortars in comparison with a reference mortar is shown.

The incorporation of wastes influenced the mortars LCA in several aspects. First, wastes incorporated as aggregates or as fillers have usually to be resized before their incorporation into mortars. That is because, in most of the cases, as described in the Chapter 3, the wastes that are generated in several industries as ceramic industry or in demolition activities do not have the suitable size to be directly applied in mortars. These treatment processes have in some cases significantly influenced the EIs of the modified mortars. Second, transportation was an aspect that considerably influenced the environmental aspects. Some wastes travelled a long distance to the factory, which had a negative impact on LCA for some categories of impacts analyzed.

Concerning aggregate wastes, from this LCA assessment, it was noticed that red ceramics and CDW wastes, even passing through a treatment process, significantly improved their eco-performance when compared with a reference mortar. For all environmental categories, these wastes presented lower or similar impacts to the ones of the reference mortar. For example, GWP was reduced by 3% when red ceramics replaced sand and by 1% when it was replaced by CDW waste. ADP (energy component) was also significantly reduced by 18% and 9%, respectively, when red ceramic or CDW were incorporated. The incorporation of sanitary ware presented several decreases of EIs in ODP, AP, POCP, ADP (energy component), or in PE-NRe. However due to the treatment process, it incremented the PE-Re impact by 5%. CDW (mainly concrete), on the other hand, was the waste that presented the highest EIs, achieving in almost all categories higher EIs than a reference mortar. This is an example that shows that the incorporation of wastes is not always sustainable. These high impacts are associated to two aspects: the wastes treatment process (crushing and sieving) and transport distance. This waste has hardness similar to a compact limestone and therefore consumes more energy and resources to be crushed, which influenced and increased its EIs. On the other side, the waste comes from a high distance, requiring travel from a considerable distance to the factory. This transportation was detrimental in the LCA analysis. To take advantage of this waste, more efficient crushing and sieving equipment and more eco-friendly means of transport are necessary.

With the exception of the GFRP waste, the filler wastes considered in the case studies came from the same sources as the aggregate wastes. These filler wastes generally differed from aggregates wastes in their size, bulk density, and volume of incorporation. These powder wastes are incorporated in percentages up to 20% of sands' volume while aggregate wastes can replace the entire natural aggregate. These differences are also noticed in the LCA. The CDW mainly composed of concrete wastes incremented almost all environmental categories as an aggregate incorporation; however, as a filler incorporation, it led to lower impacts than in the reference mortar. GFRP was the filler waste, from the case studies, that presented the highest environmental benefits in comparison with a reference mortar.

Forest biomass ashes were the only waste from the case studies that replaced cement. This waste replaced cement at 15% by volume. The improvements of this replacement in terms of EIs were considerable. Global warming potential, for example, was reduced by 13%. The benefits of this incorporation would be even higher if the waste could come from a short distance to the factory. The lorry transport had to travel almost 300 km to the factory and come back to the origin. This distance influenced the LCA analysis. This example stresses the idea that the wastes should be in the same area of the factories to increment the environmental benefits.

The incorporation of textile fibers and their influence on the mortars' LCA was also analyzed in this chapter. The fiber wastes were added to the mortars at 5% of the total volume, replacing part of the initial constituents (sand and cement). The fibers' cutting process was not taken into account in the LCA as it was considered that the influence of this process was negligible by comparison with the other processes under analysis (see Section 8.3.4.). In addition, textile fibers waste came from a short distance from the factory, about 24 km. All of this contributed to significant decreases in EIs, within the categories analyzed.

In conclusion, in this chapter it was found that wastes can improve the EIs of render mortars when replacing sand or cement. In this sense, they should be used as eco-solutions. However, some aspects should be taken in account before incorporating them. The wastes treatment processes contribute to increasing several environmental categories. In the case studies presented in this book, crushing and sieving were the treatments evaluated, but there are others and their effect in the environment should be evaluated. The transport distance is also an important issue that should be taken into account in the waste's selection. Local produced wastes promote less EIs than those from further locations.

Finally, it should be mentioned that, in this LCA methodology, the impacts of wastes deposited rather than incorporated in mortar are not taken into account in LCA analysis. Therefore, the benefits of wastes incorporation should be added to those quantified in this LCA.

References

9

[1] C.B. Farinha, J. De Brito, R. Veiga, J. Lucas, Reduction of cement content in renderings with fine sanitary ware aggregates, Mater. Struct. 49 (5) (2016) 1605–1618, doi:10.1617/s11527-015-0598-2.

[2] C.R. Gagg, Cement and concrete as an engineering material: An historic appraisal and case study analysis, Eng. Fail. Anal. 40 (2014) 114–140, doi:10.1016/j.engfailanal.2014.02.004.

[3] C.B. Cheah, M. Ramli, The implementation of wood waste ash as a partial cement replacement material in the production of structural grade concrete and mortar: An overview, Resour. Conserv. Recycl. 55 (7) (2011) 669–685, doi:10.1016/j.resconrec.2011.02.002.

[4] European Cement Research Academy (ECRA), Environmental product declaration report, Portland-composite cement (CEM II) produced in Europe (2015), doi:10.1063/1.4973978.

[5] C.B. Farinha, J. Silvestre, J. De Brito, R. Veiga, Life cycle assessment of mortars with incorporation of industrial wastes, Fibers 7 (7) (2019) 1–19, doi:https://doi.org/10.3390/fib7070059.

[6] G.A. Blengini, E. Garbarino, Resources and waste management in Turin (Italy): the role of recycled aggregates in the sustainable supply mix, J. Clean. Prod. 18 (10–11) (2010) 1021–1030, doi:10.1016/j.jclepro.2010.01.027.

[7] C. Brazão Farinha, J. de Brito, R. Veiga, Incorporation of fine sanitary ware aggregates in coating mortars, Constr. Build. Mater. 83 (2015) 194–206, doi:10.1016/j.conbuildmat.2015.03.028.

[8] C. Neno, J. De Brito, R. Veiga, Using fine recycled concrete aggregate for mortar production, Mater. Res. 17 (1) (2014) 168–177, doi:10.1590/S1516-14392013005000164.

[9] E.F. Ledesma, J.R. Jiménez, J.M. Fernández, A.P. Galvín, F. Agrela, A. Barbudo, Properties of masonry mortars manufactured with fine recycled concrete aggregates, Constr. Build. Mater. 71 (2014) 289–298, doi:10.1016/j.conbuildmat.2014.08.080.

[10] R.V. Silva, J. De Brito, R.K. Dhir, Performance of cementitious renderings and masonry mortars containing recycled aggregates from construction and demolition wastes, Constr. Build. Mater. 105 (2016) 400–415, doi:10.1016/j.conbuildmat.2015.12.171.

[11] J.R. Jiménez, J. Ayuso, M. López, J.M. Fernández, J. De Brito, Use of fine recycled aggregates from ceramic waste in masonry mortar manufacturing, Constr. Build. Mater. 40 (2013) 679–690, doi:10.1016/j.conbuildmat.2012.11.036.

[12] B. Zegardlo, M. Szelag, P. Ogrodnik, Ultra-high strength concrete made with recycled aggregate from sanitary ceramic wastes – The method of production and the interfacial transition zone, Constr. Build. Mater. 122 (2016) 736–742, doi:10.1016/j.conbuildmat.2016.06.112.

[13] B. Zegardlo, M. Szelag, P. Ogrodnik, Concrete resistant to spalling made with recycled aggregate from sanitary ceramic wastes – The effect of moisture and porosity on destructive processes occurring in fire conditions, Constr. Build. Mater. 173 (2018) 58–68, doi:10.1016/j.conbuildmat.2018.04.030.

[14] P. Ogrodnik, J. Szulej, The assessment of possibility of using sanitary ceramic waste as concrete aggregate—determination of the basic material characteristics, Materials (Basel) 8 (2018), doi:10.3390/app8071205 Applied Sciences.

Eco-efficient Rendering Mortars. DOI: 10.1016/C2018-0-04147-X

[15] J. Szulej, P. Ogrodnik, B. Klimek, Zeolite tuff and recycled ceramic sanitary ware aggregate in production of concrete, Materials (Basel) 11 (Sustainability) (2019), doi:10.3390/su11061782.

[16] A. Alves, T. Vieira, J. De Brito, J. Correia, Mechanical properties of structural concrete with fine recycled ceramic aggregates, Constr. Build. Mater. 64 (2014) 103–113, doi:10.1016/j.conbuildmat.2014.04.037.

[17] T. Vieira, A. Alves, J. De Brito, J. Correia, R.V. Silva, Durability-related performance of concrete containing fine recycled aggregates from crushed bricks and sanitary ware, Mater. Des. 90 (2016) 767–776, doi:10.1016/j.matdes.2015.11.023.

[18] J. Lucas, J. De Brito, R. Veiga, C.B. Farinha, The effect of using sanitary ware as aggregates on rendering mortars' performance, Mater. Des. 91 (2016), doi:10.1016/j.matdes.2015.11.086.

[19] D.H. de Almeida, R.H.F. Grillo, S.C. Maestrelli, C. Del Roveri, Properties of concrete manufactured with use of ceramic sanitary ware waste as aggregate, Rev. Matéria 24 (2) (2019).

[20] A. Halicka, P. Ogrodnik, B. Zegardlo, Using ceramic sanitary ware waste as concrete aggregate, Constr. Build. Mater. 48 (2013) 295–305, doi:10.1016/j.conbuildmat.2013.06.063.

[21] P. Ogrodnik and J. Szulej, Utilization of ceramic waste by using it as special concrete aggregate, 2nd International Conference on the Sustainable Energy and Environmental Development (2019), doi: 10.1088/1755-1315/214/1/012064.

[22] C. Medina, M.I.S. de Rojas, M. Frías, Reuse of sanitary ceramic wastes as coarse aggregate in eco-efficient concretes, Cem. Concr. Compos. 34 (1) (2012) 48–54, doi:10.1016/j.cemconcomp.2011.08.015.

[23] C. Medina, M.I.S. de Rojas, C. Thomas, J.A. Polanco, M. Frías, Durability of recycled concrete made with recycled ceramic sanitary ware aggregate. Inter-indicator relationships, Constr. Build. Mater. 105 (2016) 480–486, doi:10.1016/j.conbuildmat.2015.12.176.

[24] C. Medina, M. Frías, M.I.S. de Rojas, C. Thomas, J.A. Polanco, Gas permeability in concrete containing recycled ceramic sanitary ware aggregate, Constr. Build. Mater. 37 (2012) 597–605, doi:10.1016/j.conbuildmat.2012.08.023.

[25] C. Medina, M. Frías, M.I.S. De Rojas, Microstructure and properties of recycled concretes using ceramic sanitary ware industry waste as coarse aggregate, Constr. Build. Mater. 31 (2012) 112–118, doi:10.1016/j.conbuildmat.2011.12.075.

[26] C. Medina, M.I.S. de Rojas, M. Frías, Properties of recycled ceramic aggregate concretes: Water resistance, Cem. Concr. Compos. 40 (2013) 21–29, doi:10.1016/j.cemconcomp.2013.04.005.

[27] C. Medina, P.F.G. Banfill, M.I.S. de Rojas, M. Frías, Rheological and calorimetric behaviour of cements blended with containing ceramic sanitary ware and construction / demolition waste, Constr. Build. Mater. 40 (2013) 822–831, doi:10.1016/j.conbuildmat.2012.11.112.

[28] P. Torkittikul, A. Chaipanich, Utilization of ceramic waste as fine aggregate within Portland cement and fly ash concretes, Cem. Concr. Compos. 32 (6) (2010) 440–449, doi:10.1016/j.cemconcomp.2010.02.004.

[29] W. Jackiewicz-Rek, K. Załęgowski, A. Garbacz, B. Bissonnette, Properties of cement mortars modified with ceramic waste fillers, Procedia Eng., vol. 108, no. 7th Scientific-Technical Conference Material Problems in Civil Engineering (MATBUD'2015) Properties (2015) 681–687, doi:10.1016/j.proeng.2015.06.199.

[30] S. Siddique, S. Chaudhary, S. Shrivastava, T. Gupta, Sustainable utilisation of ceramic waste in concrete: Exposure to adverse conditions, J. Clean. Prod. 210 (2019) 246–255, doi:10.1016/j.jclepro.2018.10.231.

[31] S.V. Zito, E.F. Irassar, V.F. Rahhal, Management of sanitary ware wastes as supplementary cementing materials in concretes, J. Sustain. Cem. Mater. (2019), doi:10.1080/21650373.2019.1673258.

[32] A.E. Lavat, M.A. Trezza, M. Poggi, Characterization of ceramic roof tile wastes as pozzolanic admixture, Waste Manag 29 (5) (2009) 1666–1674, doi:10.1016/j.wasman.2008.10.019.

[33] M.C.S. Nepomuceno, R.A.S. Isidoro, J.P.G. Catarino, Mechanical performance evaluation of concrete made with recycled ceramic coarse aggregates from industrial brick waste, Constr. Build. Mater. 165 (2018) 284–294, doi:10.1016/j.conbuildmat.2018.01.052.

[34] A.S. El-dieb, D.M. Kanaan, Ceramic waste powder an alternative cement replacement – Characterization and evaluation, Sustain. Mater. Technol. 17 (2018), doi:10.1016/j.susmat.2018.e00063.

[35] Z. Keshavarz, D. Mostofinejad, Porcelain and red ceramic wastes used as replacements for coarse aggregate in concrete, Constr. Build. Mater. 195 (2019) 218–230, doi:10.1016/j.conbuildmat.2018.11.033.

[36] F. Cartuxo, J. De Brito, L. Evangelista, J.R. Jiménez, E.F. Ledesma, Rheological behaviour of concrete made with fine recycled concrete aggregates – Influence of the superplasticizer, Constr. Build. Mater. 89 (2015) 36–47, doi:10.1016/j.conbuildmat.2015.03.119.

[37] D. Pedro, J. De Brito, L. Evangelista, Mechanical characterization of high performance concrete prepared with recycled aggregates and silica fume from precast industry, J. Clean. Prod. 164 (2017) 939–949, doi:10.1016/j.jclepro.2017.06.249.

[38] D. Pedro, J. De Brito, L. Evangelista, Structural concrete with simultaneous incorporation of fine and coarse recycled concrete aggregates: Mechanical , durability and long-term properties, Constr. Build. Mater. 154 (2017) 294–309, doi:10.1016/j.conbuildmat.2017.07.215.

[39] D. Pedro, J. De Brito, L. Evangelista, Durability performance of high-performance concrete made with recycled aggregates , fly ash and densified silica fume, Cem. Concr. Compos. 93 (2018) 63–74, doi:10.1016/j.cemconcomp.2018.07.002.

[40] H. Higashiyama, F. Yagishita, M. Sano, O. Takahashi, Compressive strength and resistance to chloride penetration of mortars using ceramic waste as fine aggregate, Constr. Build. Mater. 26 (1) (2012) 96–101, doi:10.1016/j.conbuildmat.2011.05.008.

[41] V. Abreu, L. Evangelista, J. De Brito, The effect of multi-recycling on the mechanical performance of coarse recycled aggregates concrete, Constr. Build. Mater. 188 (2018) 480–489, doi:10.1016/j.conbuildmat.2018.07.178.

[42] J. Pacheco, J. De Brito, C. Chastre, L. Evangelista, Experimental investigation on the variability of the main mechanical properties of concrete produced with coarse recycled concrete aggregates, Constr. Build. Mater. 201 (2019) 110–120, doi:10.1016/j.conbuildmat.2018.12.200.

[43] F. Agrela, M.S. de Juan, J. Ayuso, V.L. Geraldes, J.R. Jiménez, Limiting properties in the characterisation of mixed recycled aggregates for use in the manufacture of concrete, Constr. Build. Mater. 25 (10) (2011) 3950–3955, doi:10.1016/j.conbuildmat.2011.04.027.

[44] M. Bravo, J. De Brito, L. Evangelista, J. Pacheco, Durability and shrinkage of concrete with CDW as recycled aggregates: Benefits from superplasticizer's incorporation and influence of CDW composition, Constr. Build. Mater. 168 (2018) 818–830, doi:10.1016/j.conbuildmat.2018.02.176.

[45] R.V. Silva, J. De Brito, R.K. Dhir, Tensile strength behaviour of recycled aggregate concrete, Constr. Build. Mater. 83 (2015) 108–118, doi:10.1016/j.conbuildmat.2015.03.034.

[46] R.V. Silva, J. De Brito, R.K. Dhir, Fresh-state performance of recycled aggregate concrete: A review, Constr. Build. Mater. 178 (2018) 19–31, doi:10.1016/j.conbuildmat.2018.05.149.

[47] H. Higashiyama, K. Yamauchi, M. Sappakittipakorn, M. Sano, O. Takahashi, A visual investigation on chloride ingress into ceramic waste aggregate mortars having different water to cement ratios, Constr. Build. Mater. 40 (2013) 1021–1028, doi:10.1016/j.conbuildmat.2012.11.078.

[48] H. Higashiyama, M. Sappakittipakorn, M. Sano, F. Yagishita, Chloride ion penetration into mortar containing ceramic waste aggregate, Constr. Build. Mater. 33 (2012) 48–54, doi:10.1016/j.conbuildmat.2012.01.018.

[49] J. Silva, J. De Brito, R. Veiga, Incorporation of fine ceramics in mortars, Constr. Build. Mater. 23 (1) (2009) 556–564, doi:10.1016/j.conbuildmat.2007.10.014.

[50] J. Silva, J. De Brito, R. Veiga, Fine ceramics replacing cement in mortars partial replacement of cement with fine ceramics in rendering mortars, Mater. Struct. Constr. 41 (8) (2008) 1333–1344, doi:10.1617/s11527-007-9332-z.

[51] M. Braga, J. De Brito, R. Veiga, Incorporation of fine concrete aggregates in mortars, Constr. Build. Mater. 36 (2012) 960–968, doi:10.1016/j.conbuildmat.2012.06.031.

[52] A. Aattache, R. Soltani, A. Mahi, Investigations for properties improvement of recycled PE polymer particles-reinforced mortars for repair practice, Constr. Build. Mater. 146 (2017) 603–614, doi:10.1016/j.conbuildmat.2017.04.110.

[53] L. Tosti, A. Van Zomeren, J.R. Pels, R.N.J. Comans, Technical and environmental performance of lower carbon footprint cement mortars containing biomass fly ash as a secondary cementitious material, Resour. Conserv. Recycl. Recycl. 134 (2018) 25–33, doi:10.1016/j.resconrec.2018.03.004 January.

[54] T. Ramos, A. Mafalda, J. Sousa-Coutinho, Mortar with wood waste ash: Mechanical strength carbonation resistance and ASR expansion, Constr. Build. Mater. 49 (2013) 343–351, doi:10.1016/j.conbuildmat.2013.08.026.

[55] R. Rajamma, et al., Biomass fly ash effect on fresh and hardened state properties of cement based materials, Compos. Part B Eng. 77 (2015) 1–9, doi:10.1016/j.compositesb.2015.03.019.

[56] I. Martínez-Lage, et al., Concretes and mortars with waste paper industry: Biomass ash and dregs, J. Environ. Manage. 181 (2016) 863–873, doi:10.1016/j.jenvman.2016.06.052.

[57] E.R. Teixeira, R. Mateus, A. Camões, F.G. Branco, Quality and durability properties and life-cycle assessment of high volume biomass fly ash mortar, Constr. Build. Mater. 197 (2019) 195–207, doi:10.1016/j.conbuildmat.2018.11.173.

[58] J. Usman, A. Rahman, M. Sam, S.R. Sumadi, Strength development and porosity of blended cement mortar : Effect of palm oil fuel ash content, Sustain. Environ. Res. 25 (1) (2015) 47–52.

[59] C.L. Hwang, V.A. Tran, J.W. Hong, Y.C. Hsieh, Effects of short coconut fiber on the mechanical properties, plastic cracking behavior, and impact resistance of cementitious composites, Constr. Build. Mater. 127 (2016) 984–992, doi:10.1016/j.conbuildmat.2016.09.118.

[60] A.N. Raut, C.P. Gomez, Thermal and mechanical performance of oil palm fiber reinforced mortar utilizing palm oil fly ash as a complementary binder, Constr. Build. Mater. 126 (2016) 476–483, doi:10.1016/j.conbuildmat.2016.09.034.

[61] V. Sata, J. Tangpagasit, C. Jaturapitakkul, P. Chindaprasirt, Effect of W/B ratios on pozzolanic reaction of biomass ashes in Portland cement matrix, Cem. Concr. Compos. 34 (1) (2012) 94–100, doi:10.1016/j.cemconcomp.2011.09.003.

[62] R.N. González-Kunz, P. Pineda, A. Bras, L. Morillas, Plant biomass ashes in cement-based building materials. Feasibility as eco-efficient structural mortars and grouts, Sustain. Cities Soc. 31 (2017) 151–172, doi:10.1016/j.scs.2017.03.001.

[63] J. Rosales, M. Cabrera, M.G. Beltrán, M. Lopez, F. Agrela, Effects of treatments on biomass bottom ash applied to the manufacture of cement mortars, J. Clean. Prod. 154 (2017) (2020) 424–435, doi:10.1016/j.jclepro.2017.04.024.

[64] M.G. Beltrán, A. Barbudo, F. Agrela, J.R. Jiménez, J. de Brito, Mechanical performance of bedding mortars made with olive biomass bottom ash, Constr. Build. Mater. 112 (2016) 699–707, doi:10.1016/j.conbuildmat.2016.02.065.

[65] F. Tittarelli, S.P. Shah, Effect of low dosages of waste GRP dust on fresh and hardened properties of mortars: Part 1," Constr, Build. Mater., vol. 47 (2013) 1532–1538, doi:https://doi.org/10.1016/j.conbuildmat.2013.06.043.

[66] F. Tittarelli, G. Moriconi, Use of GRP industrial by-products in cement based composites, Cem. Concr. Compos. 32 (2010) 219–225, doi:https://doi.org/10.1016/j.cemconcomp.2009.11.005.

[67] F. Tittarelli, Effect of low dosages of waste GRP dust on fresh and hardened properties of mortars: Part 2," Constr. Build. Mater., vol. 47 (2013) 1539–1543, doi:10.1016/j.conbuildmat.2013.06.086.

[68] G. Ramakrishna, T. Sundararajan, Impact strength of a few natural fibre reinforced cement mortar slabs: A comparative study, Cem. Concr. Compos. 27 (5) (2005) 547–553, doi:10.1016/j.cemconcomp.2004.09.006.

[69] S. Vantadori, A. Carpinteri, A. Zanichelli, Lightweight construction materials: Mortar reinforced with date-palm mesh fibres, Theor. Appl. Fract. Mech. 100 (2019) 39–45, doi:10.1016/j.tafmec.2018.12.011.

[70] N. Sathiparan, M. Nishanthana, B.H.M. Pavithra, Performance of coconut coir reinforced hydraulic cement mortar for surface plastering application, Constr. Build. Mater. 142 (2017) 23–30, doi:10.1016/j.conbuildmat.2017.03.058.

[71] R. Fujiyama, F. Darwish, and M. Pereira, Mechanical characterization of sisal fiber reinforced cement mortar, 13th International Conference on Fracture (2013).

[72] G. Di Bella, V. Fiore, G. Galtieri, C. Borsellino, A. Valenza, Effects of natural fibres reinforcement in lime plasters (kenaf and sisal vs . Polypropylene, Constr. Build. Mater. 58 (2014) 159–165, doi:10.1016/j.conbuildmat.2014.02.026.

[73] S. Spadea, I. Farina, A. Carrafiello, F. Fraternali, Recycled nylon fibers as cement mortar reinforcement, Constr. Build. Mater. 80 (2015) 200–209, doi:10.1016/j.conbuildmat.2015.01.075.

[74] S. Orasutthikul, D. Unno, H. Yokota, Effectiveness of recycled nylon fiber from waste fishing net with respect to fiber reinforced mortar, Constr. Build. Mater. 146 (2017) 594–602, doi:10.1016/j.conbuildmat.2017.04.134.

[75] R. Santos, "The influence of natural aggregates on the performance of replacement mortars for ancient buildings: the effects of mineralogy, grading and shape," PhD thesis in Civil Engineering, University of Lisbon, Instituto Superior Técnico, Lisbon, 2019.

[76] S. Real, Durability and thermal performance of structural concretes of light aggregates produced with different types of cementitious materials (in Portuguese), PhD Thesis, Civil Engineering, University of Lisbon, Instituto Superior Técnico, Lisbon, 2019.

[77] R. Veiga, Behaviour of Renderings Mortars: Contribution for the Research in Cracking Resistance (in Portuguese), PhD thesis in Civil Engineering, Faculty of Engineering of the University of Porto, Porto, 1998.

[78] R. Veiga, Air lime mortars: What else do we need to know to apply them in conservation and rehabilitation interventions? A review, Constr. Build. Mater. 157 (2017) 132–140, doi:10.1016/j.conbuildmat.2017.09.080.

[79] P. Faria, Coating for ancient masonry: contribution for the research in the influence of binders (in Portuguese), PhD thesis in Civil Engineering, University NOVA of Lisbon, NOVA School of Science and Technology, Lisbon, 2004.

[80] R. Santos, R. Veiga, L. Matias, A. Santos Silva, J. de Brito, Durability and compatibility of lime-based mortars: The effect of aggregates, Infrastructures 3 (3) (2019), doi:10.3390/infrastructures3030034.

[81] R. Oliveira, J. De Brito, R. Veiga, Reduction of the cement content in rendering mortars with fine glass aggregates, J. Clean. Prod. 95 (2015) 75–88, doi:10.1016/j.jclepro.2015.02.049.

[82] A. de Sousa-Coutinho, Manufacture and properties of concrete, volume I, in Portuguese, 1988.

[83] R. Hendrickx, The Adequate Measurement of the Workability of Masonry Mortar, PhD thesis in Engineering Sciences, Faculty of Engineering, Arenberg Doctoral School of Science (2009) Leuven.

[84] R. Santos, R. Veiga, A. Santos Silva, J. de Brito, Impact of aggregates on fresh mortars' properties, 5th Historic Mortars Conference (HMC), 2019.

[85] K. Terheiden, Simultaneous measurement of vapor and liquid moisture transport in porous building materials, Build. Environ. 43 (2008) 2188–2192, doi:https://doi.org/10.1016/j.buildenv.2008.01.001.

[86] J. Grilo, P. Faria, R. Veiga, A.Santos Silva, V. Silva, A. Velosa, New natural hydraulic lime mortars - Physical and microstructural properties in different curing conditions, Constr. Build. Mater. 54 (2014) 378–384, doi:10.1016/j.conbuildmat.2013.12.078.

[87] R. Veiga, M. Stefanidou, K. Balksten, J.I. Alvarez, C. Nunes, M. Theodoridou, P. Faria, I. Papayianni, R.V. Hees, Durability of lime based renders: A review of some degradation mechanisms and assessment test methods, SMS19 Rovinj (2019).

[88] R. Veiga, A. Velosa, A. Magalhães, Evaluation of mechanical compatibility of renders to apply on old walls based on a restrained shrinkage test, Materials and Structures 40 (2007) 1115–1126, doi:10.1617/s11527-006-9209-6.

[89] R. Veiga, A.Santos Silva, Long-term performance and durability of masonry structures. Degradation mechanisms, health monitoring and service life design, Part 6 - Mortars, 2019.

[90] EN 1015-21, Methods of test for mortar for masonry - Part 21: Determination of the compatibility of one-coat rendering mortars with substrates, Determination of the compatibility of one-coat rendering mortars with substrates, 2002.

[91] D ASTM5590, Determining the resistance of paint films and related coatings to fungal defacement by accelerated four-week agar plate assay, American Standard: American Society for Testing Materials (ASTM) (2000).

[92] EN 1015-3, *Methods of Test for Mortar for Masonry - Part 3: Determination of Consistence of Fresh Mortar (by Flow Table)*, European Standard: European Committee for Standardization (CEN), 1999.

[93] EN 459-2, *Building Lime. Part 2 - Test Methods*, European Standard: European Committee for Standardization (CEN), 2016.

[94] EN 413-2, *Masonry Cement. Part 2 - Test Methods*, European Committee for Standardization (CEN), 2016.

[95] BS 4551-1, *Methods of Testing Mortars, Screeds and Plasters. Physical Testing*, British Standard: British Standard Institution (BSI), 1998.

[96] ASTM C230/C30M-14, *Flow Table for Use in Tests of Hydraulic Cement*, American Standard: American Society for Testing Materials (ASTM), 2014.

[97] ASTM C1437-15, *Flow of Hydraulic Cement Mortar*, American Standard: American Society for Testing Materials (ASTM), 2015.

[98] EN 1015-4, *Methods of Test for Mortar for Masonry - Part 4: Determination of Consistence of Fresh Mortar (by Plunger Penetration)*, European Standard: European Committee for Standardization (CEN), 1998.

[99] DIN 4211:1995-03, *Masonry Mortar - Specifications, Control*, Deutsch Standard: Deutsches Institut für Normung (DIN), 1998.

[100] ASTM C187-16, *Amount of Water Required for Normal Consistency of Hydraulic Cement Paste*, American Standard: American Society for Testing Materials (ASTM), 2016.

[101] EN 1015-18, *Methods of Test for Mortar for Masonry - Part 18: Determination of Water Absorption Coefficient due to Capillary Action of Hardened Mortar*, European Standard: European Committee for Standardization (CEN), 2002.

[102] ASTM C1430-15, *Rate of Water Absorption of Masonry Mortars*, American Standard: American Society for Testing Materials (ASTM), 2015.

[103] UNI 10859, *Cultural Heritage - Natural and Artificial Stones - Determination of Water Absorption by Capillarity*, Italian Standard: Ente Nazionale Italiano di Unificazione (UNI), 2000.

[104] EN 16322, Conservation of Cultural Heritage - Test Methods - Determination of drying properties, European Standard: European Committee for Standardization (CEN), 2013.

[105] EN 1015-19, *Methods of Test for Mortar for Masonry – Part 19: Determination of Water Vapour Permeability of Hardened Rendering and Plastering Mortars*, European Standard: European Committee for Standardization (CEN), 1998.

[106] ASTM E96/E96M-16, *Water Vapor Transmission of Materials*, American Standard: American Society for Testing Materials (ASTM), 2016.

[107] EN 1015-12, *Methods of Test for Mortar for Masonry - Part 12: Determination of Adhesive Strength of Hardened Rendering and Plastering Mortars on Substrates*, European Standard: European Committee for Standardization (CEN), 2000.

[108] EN 14146, *Natural Stone Tests Methods: Determination of the Dynamic Modulus of Elasticity (by Measuring the Fundamental Resonance Frequency)*, European Standard: European Committee for Standardization (CEN), 2004.

[109] EN 12504-4, *Testing Concrete in Structures - Parte 4: Determination of Ultrasonic Pulse Velocity*, European Standard: European Committee for Standardization (CEN), 2004.

[110] EN 1015-11, *Methods of Test for Mortar for Masonry - Part 11: Determination of Flexural and Compressive Strength of Hardened Mortar*, European Standard: European Committee for Standardization (CEN), 1999.

[111] ASTM C348-18, *Flexural Strength of Hydraulic-Cement Mortars*, American Standard: American Society for Testing Materials (ASTM), 2018.

[112] ASTM C1579-13, *Evaluating Plastic Shrinkage Cracking of Restrained Fiber Reinforced Concrete (Using a Steel Form Insert)*, American Srandard: American Society for Testing Materials (ASTM), 2013.

[113] ASTM C596-18, *Drying Shrinkage of Mortar Containing Hydraulic Cement*, American Standard: American Society for Testing Materials (ASTM), 2018.

[114] UNI 6555, *Concrete Made with Aggregate Maximum Size 30 Mm - Hydraulic Shrinkage Determination*, Italian Standard: Ente Nazionale Italiano di Unificazione (UNI), 1973.

[115] Cahier 2669-4, *Certification CSTB Des Enduits Monocouches D'imperméabilisation, Modalités D'essais*, French Standard: Centre Scientifique et Technique du Bâtiment, 1993.

[116] ASTM C1581/C1581M-18a, *Determining Age at Cracking and Induced Tensile Stress Characteristics of Mortar and Concrete under Restrained Shrinkage*, American Standard: American Society for Testing Materials (ASTM), 2018.

[117] EN 477, *Unplasticized Polyvinlycloride (PVC-U) Profiles for the Fabrication of Windows and Doors - Determination of the Resistance to Impact of Main Profiles by Falling Mass*, European Standard: European Committee for Standardization (CEN), 1995.

[118] EN 13501-1, *Fire Classification of Construction Products and Building Elements - Part 1: Classification Using Data from Reation to Fire Tests*, European Standard: European Committee for Standardization (CEN), 2009.

[119] EN ISO 1182, *Reaction to Fire Tests for Products - Non-Combustibility Test*, European Standard: European Committee for Standardization (CEN), 2010.

[120] EN ISO 1716, *Reaction to Fire Tests for Products - Determination of the Gross Heat of Combustion (Calorific Value)*, European Standard: European Committee for Standardization (CEN), 2010.

[121] EN 13823, *Reaction to Fire Tests for Building Products - Building Products Excluding Floorings Exposed to the Thermal Attack by a Single Burning Item*, European Standard: European Committee for Standardization (CEN), 2010.

[122] EN ISO 11925-2, *Reaction to Fire Tests - Ignitability of Building Products Subjected to Direct Impingement of Flame - Part 2: Single-Flame Source Test*, European Standard: European Committee for Standardization (CEN), 2010.

[123] European Parliament and of the Council and Official Journal of the European Union, European Directive 2008/98/EC: On Waste and Repealing Certain Directives (Text with EEA Relevance) (2008).

[124] T. Zhang, E. Dieckmann, S. Song, J. Xie, Z. Yu, C. Cheeseman, Properties of magnesium silicate hydrate (M-S-H) cement mortars containing chicken feather fibres, Constr. Build. Mater. 180 (2018) 692–697, doi:10.1016/j.conbuildmat.2018.05.292.

[125] A.P. Fantilli, S. Sicardi, F. Dotti, The use of wool as fiber-reinforcement in cement-based mortar, Constr. Build. Mater. 139 (2017) 562–569, doi:10.1016/j.conbuildmat.2016.10.096.

[126] C.M. Pederneiras, R. Veiga, J. De Brito, Rendering mortars reinforced with natural sheep's wool fibers, Materials (Basel) 12 (22) (2019), doi:10.3390/ma12223648.

[127] G. Araya-letelier, F.C. Antico, M. Carrasco, P. Rojas, C.M. García-herrera, Effectiveness of new natural fibers on damage-mechanical performance of mortar, Constr. Build. Mater. 152 (2017) 672–682, doi:https://doi.org/10.1016/j.conbuildmat.2017.07.072.

[128] D. Cree, P. Pliya, Effect of elevated temperature on eggshell, eggshell powder and eggshell powder mortars for masonry applications, J. Build. Eng. 26 (2019), doi:10.1016/j.jobe.2019.100852.

[129] P. Pliya, D. Cree, Limestone derived eggshell powder as a replacement in Portland cement mortar, Constr. Build. Mater. 95 (2015) 1–9, doi:10.1016/j.conbuildmat.2015.07.103.

[130] C. Martínez-garcía, B. González-fonteboa, F. Martínez-abella, D.C.- López, "Virtual special issue bio based building materials performance of mussel shell as aggregate in plain concrete," vol. 139, pp. 570–583, 2017, doi: 10.1016/j.conbuildmat.2016.09.091.

[131] J.R. González-López, J.F. Ramos-Lara, A. Zaldivar-Cadena, L. Chávez-Guerrero, R.X. Magallanes-Rivera, O. Burciaga-Díaz, Small addition effect of agave biomass ashes in cement mortars, Fuel Process. Technol. 133 (2015) 35–42, doi:10.1016/j.fuproc.2014.12.041.

[132] D.K. Panesar, B. Shindman, The mechanical, transport and thermal properties of mortar and concrete containing waste cork, Cem. Concr. Compos. 34 (9) (2012) 982–992, doi:10.1016/j.cemconcomp.2012.06.003.

[133] B.C. Thomas, Y.S. Jose, Impact of sisal fiber reinforced concrete and its performance analysis: a review, Evol. Intell., no. Special Issue (2019), doi:10.1007/s12065-019-00230-9.

[134] X. Xie, Z. Zhou, M. Jiang, X. Xu, Z. Wang, D. Hui, Cellulosic fibers from rice straw and bamboo used as reinforcement of cement-based composites for remarkably improving mechanical properties, Compos. Part B Eng. 78 (2015) 153–161, doi:10.1016/j.compositesb.2015.03.086.

[135] H. Savastano, P.G. Warden, R.S.P. Coutts, Brazilian waste fibres as reinforcement for cement-based composites, Cem. Concr. Compos. 22 (5) (2000) 379–384, doi:10.1016/S0958-9465(00)00034-2.

[136] N. Saikia, J. De Brito, Use of plastic waste as aggregate in cement mortar and concrete preparation: A review, Constr. Build. Mater. 34 (2012) 385–401, doi:10.1016/j.conbuildmat.2012.02.066.

[137] A. Silva, J. De Brito, R. Veiga, Incorporation of fine plastic aggregates in rendering mortars, Constr. Build. Mater. 71 (2014) 226–236, doi:https://doi.org/10.1016/j.conbuildmat.2014.08.026.

[138] B. Coppola, L. Courard, F. Michel, L. Incarnato, L. Di Maio, Investigation on the use of foamed plastic waste as natural aggregates replacement in lightweight mortar, Compos. Part B 99 (2016) 75–83, doi:10.1016/j.compositesb.2016.05.058.

[139] O.Y. Marzouk, R.M. Dheilly, M. Queneudec, Valorization of post-consumer waste plastic in cementitious concrete composites, Waste Manag 27 (2007) 310–318, doi:10.1016/j.wasman.2006.03.012.

[140] K. Hannawi, S. Kamali-Bernard, W. Prince, Physical and mechanical properties of mortars containing PET and PC waste aggregates, Waste Manag 30 (11) (2010) 2312–2320, doi:10.1016/j.wasman.2010.03.028.

[141] V. Corinaldesi, A. Nardinocchi, J. Donnini, Reuse of recycled glass in mortar manufacturing, Eur. J. Environ. Civ. Eng. 20 (2016) s140–s151, doi:10.1080/19648189.2016.1246695 October.

[142] R. Oliveira, J. De Brito, R. Veiga, Incorporation of fine glass aggregates in renderings, Constr. Build. Mater. 44 (2013) 329–341, doi:10.1016/j.conbuildmat.2013.03.042.

[143] P. Penacho, J. De Brito, R. Veiga, Physico-mechanical and performance characterization of mortars incorporating fine glass waste aggregate, Cem. Concr. Compos. 50 (2014) 47–59, doi:10.1016/j.cemconcomp.2014.02.007.

[144] G. Araya-Letelier, P. Maturana, M. Carrasco, F.C. Antico, M.S. Gómez, Mechanical-damage behavior of mortars reinforced with recycled polypropylene fibers, Sustainability 11 (8) (2019), doi:10.3390/su11082200.

[145] K. Bendjillali, M. Chemrouk, M.S. Goual, B. Boulekbache, Behaviour of polypropylene fibre mortars conserved in different environments, European Journal of Environmental and Civil Engineering 17 (8) (2013) 687–699, doi:10.1080/19648189.2013.812685.

[146] K. Bendjillali, M. Chemrouk, B. Boulekbache, Performances of cementitious mortars containing recycled synthetic fibres under hot-dry climate, Eur. J. Environ. Civ. Eng. 23 (2019) 1235–1247, doi:10.1080/19648189.2017.1344152.

[147] M.C. Bignozzi, A. Saccani, F. Sandrolini, New polymer mortars containing polymeric wastes. Part 1. Dynamic mechanical and dielectric behaviour, Compos. - Part A Appl. Sci. Manuf. 33 (2) (2002) 205–211, doi:10.1016/S1359-835X(01)00093-8.

[148] D. Pedro, J. De Brito, R. Veiga, Mortars made with fine granulate from shredded tires, J. Mater. Civ. Eng. 25 (4) (2013) 519–529, doi:10.1061/(ASCE)MT.1943-5533.0000606.

[149] L. Gu, T. Ozbakkaloglu, Use of recycled plastics in concrete: A critical review, Waste Manag 51 (2016) 19–42, doi:10.1016/j.wasman.2016.03.005.

[150] R. Siddique, J. Khatib, I. Kaur, Use of recycled plastic in concrete: A review, Waste Manag 28 (2008) 1835–1852, doi:10.1016/j.wasman.2007.09.011.

[151] F. Pelisser, N. Zavarise, T. Arent, A. Michael, Concrete made with recycled tire rubber: Effect of alkaline activation and silica fume addition, J. Clean. Prod. 19 (6–7) (2011) 757–763, doi:10.1016/j.jclepro.2010.11.014.

[152] S. Samarakoon, P. Ruben, J.W. Pedersen, L. Evangelista, Mechanical performance of concrete made of steel fibers from tire waste, Case Studies in Construction Materials 11 (2019), doi:10.1016/j.cscm.2019.e00259.

[153] H. Fathi, T. Lameie, M. Maleki, R. Yazdani, Simultaneous effects of fiber and glass on the mechanical properties of self-compacting concrete, Constr. Build. Mater. 133 (2017) 443–449, doi:10.1016/j.conbuildmat.2016.12.097.

[154] S. Jesus, C. Maia, C.B. Farinha, J. De Brito, R. Veiga, Rendering mortars with incorporation of very fine aggregates from construction and demolition waste, Constr. Build. Mater. 229 (2019), doi:10.1016/j.conbuildmat.2019.116844.

[155] S. Roque, C.M. Pederneiras, C. Brazão Farinha, J. De Brito, R. Veiga, Concrete-based and mixes waste aggregates in rendering mortars, Materials (Basel) 13 (8) (2020), doi:10.3390/ma13081976.

[156] T. Hoque, M.H. Rashid, R. Hasan, E.F. Mondol, Influence of stone dust as partially replacing material of cement and sand on some mechanical properties of mortar, Int. J. Adv. Struct. Geotech. Eng. 02 (02) (2013).

[157] J. Norambuena-Contreras, J. Quilodran, I. Gonzalez-Torre, M. Chavez, R. Borinaga-Treviño, Electrical and thermal characterisation of cement-based mortars containing recycled metallic waste, J. Clean. Prod. 190 (2018) 737–751, doi:10.1016/j.jclepro.2018.04.176.

[158] H. Nguyen, V. Carvelli, T. Fujii, K. Okubo, Cement mortar reinforced with reclaimed carbon fibres, CFRP waste or prepreg carbon waste, Constr. Build. Mater. 126 (2016) 321–331, doi:10.1016/j.conbuildmat.2016.09.044.

[159] E. Zornoza, P. Garcés, J. Payá, M.A. Climent, Improvement of the chloride ingress resistance of OPC mortars by using spent cracking catalyst, Cem. Concr. Res. 39 (2009) 126–139, doi:10.1016/j.cemconres.2008.11.006.

[160] B. Pacewska, I. Wilinska, M. Bukowska, W. Nocun-Wczelik, Effect of waste aluminosilicate material on cement hydration and properties of cement mortars, Cem. Concr. Res. 32 (2002) 1823–1830, doi:https://doi.org/10.1016/S0008-8846(02)00873-6.

[161] N. Su, H.Y. Fang, Z.H. Chen, F.S. Liu, Reuse of waste catalysts from petrochemical industries for cement substitution, Cem. Concr. Res. 30 (2000) 1773–1783, doi:10.1016/S0008-8846(00)00401-4.

[162] C. Costa, R. Geada, Effect of the incorporation of the oil industry waste on the mechanical properties of mortars (in Portuguese), 4th Portuguese Congress of Mortars and ETICS (2012).

[163] C. Costa, M.S. Ribeiro, N. Brito, Effect of waste oil-cracking catalyst incorporation on durability of mortars, Mater. Sci. Appl. 5 (2014) 905–914, doi:10.4236/msa.2014.513092.

[164] S. Nunes, C. Costa, Numerical optimization of self-compacting mortar mixture containing spent equilibrium catalyst from oil refinery, J. Clean. Prod. 158 (2017) 109–121, doi:10.1016/j.jclepro.2017.04.161.

[165] C. Costa, S. Ribeiro, N. Brito, Concrete repair mortar made of a waste from oil refining industry short summary, International Sustainable Building Conference (2013) 1455–1462.

[166] E. Zornoza, J. Payá, J. Monzó, M.V. Borrachero, P. Garcés, The carbonation of OPC mortars partially substituted with spent fluid catalytic catalyst (FC3R) and its influence on their mechanical properties, Constr. Build. Mater. 23 (2009) 1323–1328, doi:10.1016/j.conbuildmat.2008.07.024.

[167] R. de Oliveira, Effect of the incorporation of an oil refining waste into industrial mortars for steel reinforcement anchor (in Portuguese), Master Thesis in Civil Engineering (2018).

[168] J. Payá, J. Monzó, M.V. Borrachero, Fluid catalytic cracking catalyst residue (FC3R): An excellent mineral by-product for improving early-strength development of cement mixtures, Cem. Concr. Res. 29 (11) (1999) 1773–1779, doi:10.1016/S0008-8846(99)00164-7.

[169] Y. Tseng, C. Huang, K. Hsu, The pozzolanic activity of a calcined waste FCC catalyst and its effect on the compressive strength of cementitious materials, Cem. Concr. Res. 35 (2005) 782–787, doi:10.1016/j.cemconres.2004.04.026.

[170] V. Ferrándiz-Mas, E. García-Alcocel, Durability of expanded polystyrene mortars, Constr. Build. Mater. 46 (2013) 175–182, doi:10.1016/j.conbuildmat.2013.04.029.

[171] F. Tittarelli, C. Giosuè, A. Mobili, di C. Perna, S. Monosi, Effect of using recycled instead of virgin EPS in lightweight mortars, Procedia Eng 161 (2016) 660–665, doi:10.1016/j.proeng.2016.08.728.

[172] C.B. Farinha, J. De Brito, R. Veiga, Assessment of glass fibre reinforced polymer waste reuse as filler in mortars, J. Clean. Prod. 210 (2019) 1579–1594, doi:10.1016/j.jclepro.2018.11.080.

[173] M. Criado, I. García-Díaz, J.M. Bastidas, F.J. Alguacil, F.A. López, C. Monticelli, Effect of recycled glass fiber on the corrosion behavior of reinforced mortar, Constr. Build. Mater. 64 (2014) (2014) 261–269, doi:10.1016/j.conbuildmat.2014.04.049.

[174] H. Rodin, S. Nassiri, K. Englund, O. Fakron, H. Li, Recycled glass fiber reinforced polymer composites incorporated in mortar for improved mechanical performance, Constr. Build. Mater. 187 (2018) 738–751, doi:10.1016/j.conbuildmat.2018.07.169.

[175] C.B. Farinha, J. de Brito, R. Veiga, J.M. Fernández, J.R. Jiménez, A.R. Esquinas, Wastes as aggregates, binders or additions in mortars: Selecting their role based on characterization, Materials 11 (3) (2018), doi:10.3390/ma11030453.

[176] J. Pinto, et al., Render reinforced with textile threads, Constr. Build. Mater. 40 (2013) 26–32, doi:10.1016/j.conbuildmat.2012.09.099.

[177] M. Barbero-barrera, O. Pombo, M.D.L. Angeles, Textile fibre waste bindered with natural hydraulic lime, Compos. Part B 94 (2016) 26–33, doi:10.1016/j.compositesb.2016.03.013.

[178] C. Gonilho-Pereira, P. Faria, R. Fangueiro, A. Martins, P. Vinagre, Performance Assessment of Waste Fiber-Reinforced Mortar, Mater. Sci. Forum (2013) 617–622, doi:10.4028/www.scientific.net/MSF.730-732.617.

[179] M.D.L. Garcia, J. Sousa-Coutinho, Strength and durability of cement with forest waste bottom ash, Constr. Build. Mater. 41 (2013) 897–910, doi:10.1016/j.conbuildmat.2012.11.081.

[180] R.C.E. Modolo, V.M. Ferreira, L.A. Tarelho, J.A. Labrincha, L. Senff, L. Silva, Mortar formulations with bottom ash from biomass combustion, Constr. Build. Mater. 45 (2013) 275–281, doi:10.1016/j.conbuildmat.2013.03.093.

[181] T. Ramos, A.M. Matos, J. Sousa-coutinho, Strength and durability of mortar using cork waste ash as cement replacement, Mater. Res. 17 (4) (2014) 893–907, doi:https://doi.org/10.1590/S1516-14392014005000092.

[182] R. Rajamma, R.J. Ball, L.A.C. Tarelho, G.C. Allen, J.A. Labrincha, V.M. Ferreira, Characterisation and use of biomass fly ash in cement-based materials, J. Hazard. Mater. 172 (2–3) (2009) 1049–1060, doi:10.1016/j.jhazmat.2009.07.109.

[183] N. Almeida, F. Branco, R. Santos, Recycling of stone slurry in industrial activities: Application to concrete mixtures, Build. Environ. 42 (2007) 810–819, doi:10.1016/j.buildenv.2005.09.018.

[184] N. Almeida, F. Branco, J. De Brito, J.R. Santos, High-performance concrete with recycled stone slurry, Cem. Concr. Res. 37 (2006) 210–220, doi:10.1016/j.cemconres.2006.11.003.

[185] B. Pacewska, M. Bukowska, I. Wilinska, M. Swat, Modification of the properties of concrete by a new pozzolan: A waste catalyst from the catalytic process in a fluidized bed, Cem. Concr. Res. 32 (2002) 145–152.

[186] Y. Morozov, A.S. Castela, A.P.S. Dias, M.F. Montemor, Chloride-induced corrosion behavior of reinforcing steel in spent fluid cracking catalyst modified mortars, Cem. Concr. Res. 47 (2013) 1–7, doi:10.1016/j.cemconres.2013.01.011.

[187] R. Neves, C. Vicente, A. Castela, M.F. Montemor, Durability performance of concrete incorporating spent fluid cracking catalyst, Cem. Concr. Compos. 55 (2015) 308–314, doi:10.1016/j.cemconcomp.2014.09.018.

[188] S.Y.N.N. Chan, X. Ji, Comparative study of the initial surface absorption and chloride diffusion of high performance zeolite, silica fume and PFA concretes, Cem. Concr. Compos. 21 (4) (1999) 293–300, doi:10.1016/S0958-9465(99)00010-4.

[189] D. García, I. Vegas, I. Cacho, Mechanical recycling of GFRP waste as short-fiber reinforcements in microconcrete, Constr. Build. Mater. 64 (2014) 293–300, doi:10.1016/j.conbuildmat.2014.02.068.

[190] A. Dehghan, K. Peterson, A. Shvarzman, Recycled glass fiber reinforced polymer additions to Portland cement concrete, Constr. Build. Mater. 146 (2017) 238–250, doi:10.1016/j.conbuildmat.2017.04.011.

[191] A. Yazdanbakhsh, L.C. Bank, K. Rieder, Y. Tian, C. Chen, Concrete with discrete slender elements from mechanically recycled wind turbine blades, Resour. Conserv. Recycl. Recycl. 128 (2018) 11–21, doi:10.1016/j.resconrec.2017.08.005.

[192] J.R. Correia, N.M. Almeida, J.R. Figueira, Recycling of FRP composites: Reusing fine GFRP waste in concrete mixtures, J. Clean. Prod. 19 (15) (2011) 1745–1753, doi:10.1016/j.jclepro.2011.05.018.

[193] P. Asokan, M. Osmani, A.D.F. Price, Assessing the recycling potential of glass fibre reinforced plastic waste in concrete and cement composites, J. Clean. Prod. 17 (9) (2009) 821–829, doi:10.1016/j.jclepro.2008.12.004.

[194] P. Asokan, M. Osmani, A. Price, Improvement of the mechanical properties of glass fibre reinforced plastic waste powder filled concrete, Constr. Build. Mater. 24 (4) (2010) 448–460, doi:10.1016/j.conbuildmat.2009.10.017.

[195] K. Ogi, T. Shinoda, M. Mizui, Strength in concrete reinforced with recycled CFRP pieces, Compos. - Part A Appl. Sci. Manuf. 36 (2005) 893–902, doi:10.1016/j.compositesa.2004.12.009.

[196] A.U. Elinwa, Y.A. Mahmood, Ash from timber waste as cement replacement material, Cem. Concr. Compos. 24 (2) (2002) 219–222, doi:10.1016/S0958-9465(01)00039-7.

[197] M. Berra, T. Mangialardi, A. Evangelista, Reuse of woody biomass fly ash in cement-based materials, Constr. Build. Mater. 76 (2015) 286–296, doi:10.1016/j.conbuildmat.2014.11.052.

[198] C.A. Juarez, G. Fajardo, S. Monroy, A. Duran-Herrera, P. Valdez, C. Magniont, Comparative study between natural and PVA fibers to reduce plastic shrinkage cracking

in cement-based composite, Constr. Build. Mater. 91 (2015) 164–170, doi:10.1016/j.conbuildmat.2015.05.028.

[199] R. Barbosa, N. Lapa, D. Dias, B. Mendes, Concretes containing biomass ashes: Mechanical, chemical, and ecotoxic performances, Constr. Build. Mater. 48 (2013) 457–463, doi:10.1016/j.conbuildmat.2013.07.031.

[200] M. Khorami, E. Ganjian, The effect of limestone powder, silica fume and fibre content on flexural behaviour of cement composite reinforced by waste Kraft pulp, Constr. Build. Mater. 46 (2013) 142–149, doi:10.1016/j.conbuildmat.2013.03.099.

[201] S. Subas, M. Emirog, Utilizing of waste ceramic powders as filler material in self-consolidating concrete, Constr. Build. Mater. 149 (2017) 567–574, doi:10.1016/j.conbuildmat.2017.05.180.

[202] M. Mastali, A. Dalvand, A. Sattarifard, The impact resistance and mechanical properties of the reinforced self-compacting concrete incorporating recycled CFRP fiber with different lengths and dosages, Compos. Part B Eng. 112 (2017) 74–92, doi:10.1016/j.compositesb.2016.12.029.

[203] V. Corinaldesi, Influence of lightweight aggregates and GRP by-product powders on the properties of self-compacting concretes, Adv. Mater. Res. 548 (2012) 215–220, doi:10.1186/1747-1028-2-9.

[204] J. Cuenca, J. Rodríguez, M. Martín-Morales, Z. Sánchez-Roldán, M. Zamorano, Effects of olive residue biomass fly ash as filler in self-compacting concrete, Constr. Build. Mater. 40 (2013) 702–709, doi:10.1016/j.conbuildmat.2012.09.101.

[205] M. Safiuddin, J.S. West, K.A. Soudki, Flowing ability of the mortars formulated from self-compacting concretes incorporating rice husk ash, Constr. Build. Mater. 25 (2) (2011) 973–978, doi:10.1016/j.conbuildmat.2010.06.084.

[206] F. Ferella, V. Innocenzi, F. Maggiore, Oil refining spent catalysts: A review of possible recycling technologies, Resources, Conserv. Recycl 108 (2016) 10–20, doi:10.1016/j.resconrec.2016.01.010.

[207] J. Payá, J. Monzó, M.V. Borrachero, S. Velázquez, Evaluation of the pozzolanic activity of fluid catalytic cracking catalyst residue (FC3R). Thermogravimetric analysis studies on FC3R-Portland cement pastes, Cem. Concr. Res. 33 (2003) 603–609, doi:https://doi.org/10.1016/S0008-8846(02)01026-8.

[208] V. Antonovič, P. Baltrėnas, M. Aleknevičius, I. Pundienė, R. Stonys, Modification of petrochemical fluid catalytic cracking catalyst waste properties by treatment in high temperature, The 10th International Conference on Mordern Building Materials, Structures and, Techniques (2010).

[209] D. Vaičiukynienė, A. Grinys, Vi. Vaitkevičius, A. Kantautas, Purified waste FCC catalyst as a cement replacement material, Ceram. Silikaty 59 (2) (2015) 103–108.

[210] R.V. Silva, J. de Brito, C.J. Lynn, R.K. Dhir, Environmental impacts of the use of bottom ashes from municipal solid waste incineration: A review, Resour. Conserv. Recycl. 140 (September 2018) 23–35, doi:10.1016/j.resconrec.2018.09.011 2019.

[211] S.-C. Kou, B.-J. Zhan, C.-S. Poon, Use of a CO2 curing step to improve the properties of concrete prepared with recycled aggregates, Cem. Concr. Compos. 45 (2014) 22–28, doi:10.1016/j.cemconcomp.2013.09.008.

[212] N. Zhang, H. Duan, T.R. Miller, V.W.Y. Tam, G. Liu, J. Zuo, Mitigation of carbon dioxide by accelerated sequestration in concrete debris, Renew. Sustain. Energy Rev. 117 (2020) (July 2019) 109495, doi:10.1016/j.rser.2019.109495.

[213] S.-C. Kou, C.-S. Poon, Properties of concrete prepared with PVA-impregnated recycled concrete aggregates, Cem. Concr. Compos. 32 (8) (2010) 649–654, doi:10.1016/j.cemconcomp.2010.05.003.

[214] W. Zeng, Y. Zhao, C.S. Poon, Z. Feng, Z. Lu, S.P. Shah, Using microbial carbonate precipitation to improve the properties of recycled aggregate, Constr. Build. Mater. 228 (2019) 116743, doi:10.1016/j.conbuildmat.2019.116743.

[215] C.B. Farinha, J. De Brito, R. Veiga, Influence of forest biomass bottom ashes on the fresh, water and mechanical behaviour of cement-based mortars, Resour. Conserv. Recycl. 149 (2019) 750–759, doi:10.1016/j.resconrec.2019.06.020 April.

[216] A. Lozano-Lunar, P. Raposeiro da Silva, J. de Brito, J.M. Fernández, J.R. Jiménez, Safe use of electric arc furnace dust as secondary raw material in self-compacting mortars production, J. Clean. Prod. 211 (2019) 1375–1388, doi:10.1016/j.jclepro.2018.12.002.

[217] N.F. P. 18-513, Pozzolanic addition for concrete - Metakaolin: Definitions, specifications and conformity criteria (in French), French Standard, 2010.

[218] C.B. Farinha, High Performance wall rendering cementitious mortars with industrial wastes, Instituto Superior Técnico. Universidade de Lisboa. PhD thesis in Civil Engineering, 2020.

[219] EN 196-5, Methods of Testing Cement—Part 5: Pozzolanicity test for pozzolanic cement, European Standard: European Committee for Standardization (CEN), 2011.

[220] EN 196-1, Methods of Testing Cement—Part 1: Determination of Strength, European Standard: European Committee for Standardization (CEN), 2016.

[221] J. Silva, J. de Brito, R. Veiga, Recycled red-clay ceramic construction and demolition waste for mortars production, J. Mater. Civ. Eng. 22 (3) (2010) 236–244, doi:10.1061/(ASCE)0899-1561(2010)22:3(236).

[222] G.D. Taylor, Construction Materials (in Spanish), Longman Scientific & Technical, 1991.

[223] C. B. Farinha, C. Maia, J. De Brito, R. Veiga, Control of cracking of coating mortars through incorporation of textile fibre waste, ENCORE 2020 (2020).

[224] M. Halvaei, M. Jamshidi, H.R. Pakravan, M. Latifi, Interfacial bonding of fine aggregate concrete to low modulus fibers, Constr. Build. Mater. 95 (2015) 117–123, doi:10.1016/j.conbuildmat.2015.07.024.

[225] T. Lourenço, L. Matias, P. Faria, Anomalies detection in adhesive wall tiling systems by infrared thermography, Constr. Build. Mater. 148 (2017) 419–428, doi:10.1016/j.conbuildmat.2017.05.052.

[226] V. Peixoto de Freitas, M.I. Torres, A.S. Guimarães, Rising humidity, 1 edition, (*in Portuguese*), 2008.

[227] T. Santos, L. Nunes, P. Faria, Production of eco-efficient earth-based plasters: Influence of composition on physical performance and bio-susceptibility, J. Clean. Prod. 167 (2018) 55–67, doi:10.1016/j.jclepro.2017.08.131.

[228] M. Martins, L. Nunes, F. Branco, Plastering and rendering mortars incorporating regranulated of expanded cork to improve durability, International Conference on Durability of Building Materials and Components (2011).

[229] C.B. Farinha, L. Nunes, J. De Brito, R. Veiga, Evaluation of the influence of the incorporation of organic wastes in mortars in the development of moulds (in Portuguese), Construction 2018 (2018).

[230] J. L. Parracha, A. Cortay, G. Borsoi, R. Veiga, L. Nunes, Evaluation of ETICS characteristics that affect surface mould development, XV International Conference on Durability of Building, Materials and Components (2020).

[231] J. L. Parracha, et al., Performance parameters of ETICS: Correlating water resistance, biosusceptibility and surface properties, Constr. Build. Mater. 272 (2021), https://doi.org/10.1016/j.conbuildmat.2020.121956.

[232] J.D. Silvestre, Life cycle assessment "from cradle to cradle" of building assemblies - Application to external walls, Instituto Superior Técnico. Portugal. PhD thesis in Civil Engineering, 2012.

[233] C. Jiménez, M. Barra, A. Josa, S. Valls, LCA of recycled and conventional concretes designed using the Equivalent Mortar Volume and classic methods, Constr. Build. Mater. 84 (2015) 245–252, doi:10.1016/j.conbuildmat.2015.03.051.

[234] O. Ortiz, F. Castells, G. Sonnemann, Sustainability in the construction industry: A review of recent developments based on LCA, Constr. Build. Mater. 23 (1) (2009) 28–39, doi:10.1016/j.conbuildmat.2007.11.012.

[235] U.S. Environmental Protection Agency (EPA), "Life cycle assessment: Principles and practice," 2006.

[236] ISO 14040, International Standard, Environmental management – Life cycle assessment – Principles and Framework, 2006.

[237] ISO 14044, Environmental management - Life cycle assessment - Principles and framework. 2006.

[238] EN 15804, European Standard. Sustainability of construction works - Environmental product declarations - Core rules for the product category of construction products, European Committee for Standardization (CEN), 2012.

[239] European Life Cycle Database and European Life Cycle Database (ELCD), "European Life Cycle Database (ELCD)." 2013.

[240] European Federation of Concrete Admixtures Associations Ltd. (EFCA), Environmental product declaration report, Concrete admixtures - Air entrainers (2015).

[241] European Federation of Concrete Admixtures Associations Ltd (EFCA), Environmental product declaration report, Concrete admixtures – Plasticisers and superplasticisers (2015).

[242] European Federation of Concrete Admixtures Associations Ltd (EFCA), Environmental product declaration report, Concrete admixtures – Water resisting admixtures (2015).

[243] M. Braga, D. Silvestre, J. De Brito, Compared environmental and economic impact from cradle to gate of concrete with natural and recycled coarse aggregates, J. Clean. Prod. 162 (2017) 529–543, doi:10.1016/j.jclepro.2017.06.057.

[244] R. Kurda, J.D. Silvestre, J. De Brito, H. Ahmed, Effect of incorporation of high volume of recycled concrete aggregates and fly ash on the strength and global warming potential of concrete, J. Clean. Prod. 166 (2017) 485–502, doi:10.1016/j.jclepro.2017.07.236.

[245] R. Kurda, Sustainable development of cement-based materials: Application to recycled aggregates concrete, Instituto Superior Técnico. Universidade de Lisboa. PhD thesis in Civil Engineering (2017).

[246] G.A. Blengini, Life cycle assessment tools for sustainable development: case studies for the mining and construction industries in Italy and Portugal, Instituto Superior Técnico. Universidade de Lisboa. PhD thesis in Mining Engineering (2006).

[247] R. Veiga, Report 427/05-NRI: Rules for the Granting of Application Documents to Pre-Dosed Cement-Based Mineral Binder Coatings for Walls (in Portuguese), National Laboratory for Civil Engineering (LNEC), 2005.

Index

Page numbers followed by "*f*" and "*t*" indicate, figures and tables respectively.

Made in the USA
Monee, IL
13 September 2026

63138392R00146